AF520828

A to Z
GEOLOGY OF INDIA
(Stratigraphy and Fossils)

About the Author

O.P. Mathur did his M.Sc. in Geology from the Banaras Hindu University, Varanasi, U.P. in 1962. During 1962-63 he was lecturer in geology at M.B. College, Udaipur, Rajasthan. In 1963-64 he worked as an engineering geologist with the U.P. Irrigation Department and was engaged in the geological investigation for the construction of the Tehri Dam on the river Bhagirathi. In December 1964 the author joined the Geological Survey of India and remained engaged in geological mapping on the aerial photographs for iron ore investigation in the Sikar and Jhunjhunu districts of Rajasthan. In 1972 the author opted for the newly created Central Ground Water Board and worked for the ground water exploration in western Rajasthan and hydrogeological mapping in the Dungarpur district in south Rajasthan. From 1975 to 1978 the author worked in the UNDP Project of the Central Ground Water Board on the Ghaggar River Basin and was associated with the Artificial Recharge studies. From 1978 to 1985 the author worked in the North-Western Region of the Central Ground Water Board, Chandigarh, and conducted a number of short-term water supply investigations in addition to other ground water studies. From 1985 till his retirement in 1997 as Director, he remained as officer in charge of the state Unit Office, Western Region, Central Ground Water Board, Jodhpur, Rajasthan and supervised hydrogeological studies of the Unit Office. Presently the author residents at Udaipur.

A to Z
GEOLOGY OF INDIA
(Stratigraphy and Fossils)

(A Bedside Book)

O. P. Mathur

SCIENTIFIC
PUBLISHERS

Published by
SCIENTIFIC PUBLISHERS (INDIA)

Jodhpur –

5 A, New Pali Road
P.O. Box 91
Jodhpur - 342 001 INDIA

Delhi –

4806/24, Ansari Road
Daryaganj
New Delhi - 110 002 INDIA

Print: 2024

All rights reserved. No part of this publication or the information contained herein may be reproduced, adapted, abridged, translated, stored in a retrieval system, computer system, photographic or other systems or transmitted in any form or by any means, electronic, mechanical, optical, digital, by photocopying, recording or otherwise, without written prior permission from the publisher. Any breach will attract legal action and prosecution without further notice.

Disclaimer: While every effort has been made to avoid errors and omissions, this publication is being sold and marketed on the understanding and presumption that neither the editors (or authors) nor the publishers nor the printers would be liable in any manner whatsoever, to any person either for an error or for an omission in this publication, or for any action to be taken on the basis of this work. Any inadvertent discrepancy noted may be brought to the attention of the publisher, for rectifying it in future editions, if published.

This book contains information obtained from authentic and highly regarded sources. Reasonable efforts have been made to publish reliable data and information, but the editors and publisher cannot assume responsibility for the validity of all materials or the consequences of their use. The editors and publisher have attempted to trace and acknowledge the copyright holders of all material reproduced in this publication and apologize to copyright holders if permission and acknowledgement to publish in this form have not been obtained. If any copyright material has not been acknowledged please write and let us know so that we may rectify it.

Trademark Notice: Publications or corporate names may be trademarks, and are used only for identification and explanation in bonafide intent without intent to infringe.

ISBN: 978-93-86652-11-9
eISBN: 978-93-87869-49-3

© 2018, Scientific Publishers (India)

Visit the Scientific Publishers (India) website at
http://www.scientificpubonline.com

Printed in India

CONTENTS

I.	Preface	1
II.	List of Formations	2
III.	Description of Formations	15
IV.	List of Tables	83
V.	List of fossils with their occurrences and fossils groups	107
VI.	List of fossiliferous formations with their fossils	185
VII.	List of fossil groups pertaining to this book	192
VIII.	Details of the fossil groups	194
IX.	List of fossils as per the fossil groups	198
X.	List of fossiliferous formations	201

I

PREFACE

I am glad to present this book on the geology of India, covering stratigraphy and fossils. The book is in the form of a ready reference. The subject matter of stratigraphy is full of names of formations, groups, etc. At times when we need to know about a certain formation for which we do not know the exact stratigraphic position, then we have to search the entire book, page by page, which becomes quite irritating and time consuming. To overcome this problem, the idea of arranging the different formations in an alphabetic order occurred to the author. During this process it was seen that many small formations, which are otherwise important, do not get their due representation, because they lie in company or association with much larger formations. Otherwise also stratigraphy is nothing but an orderly and chronological arrangement of different formations. In other words, we can say that stratigraphy is a language by itself, where different formations are its words. Same is the case with large number of fossils occurring in different formations. In the usual literature on the subject, it is practically impossible to find in what formation or formations a particular fossil occurs and to which fossil group it belongs. Alphabetical arrangement of fossils as shown in the list of fossils will help students and scholars to persue their task in an easier and quicker way. To arrange such a large number of fossils in an alphabetical order and to find their fossil groups was really a tough job. Still the author does not claim that all the formations and fossils occurring in the Indian stratigraphy are included in this book, and it cannot be the last word on the subject. This is more true in case of fossils, where unlimited literature is available. However, the author feels satisfied to present the subject matter in a totally different and new manner. I am sure that even if someone goes casually or leisurely through this book, he cannot escape without adding something to his knowledge and familiarity with the subject and that is why I have called it a bedside book, in addition to a reference book. The idea and desire to write such a book was born during the author's teaching tenure, but for various reasons the same could not be presented earlier. The author wishes to state that the information presented in this book has been extracted from the existing literature on the subject which is quite voluminous and therefore, it was not possible to mention references in this small book. I again emphasize that this is only a compilation of the existing information, but the mode of presentation is different.

■■■■

II

LIST OF FORMATIONS

1. A1 Abor Volcanic Formation
2. A2 Abujhmar Basin
3. A3 Abujhmar Group
4. A4 Agglomeratic Slate
5. A5 Ahmadnagar Sandstone
6. A6 Aida Stage
7. A7 Aishmuqam Formation
8. A8 Ajabgarh Group
9. A9 Ajmer Formation
10. A10 Akli Formation
11. A11 Alalor Formation
12. A12 Albaka Pamanuru (Chelimala) Formation
13. A13 Albitite
14. A14 Alluvium, Newer
15. A15 Alluvium, Older
16. A16 Almora Gneiss
17. A17 Alwar Basin Granites
18. A18 Alwar Group
19. A19 Ambalapuzha Formation
20. A20 Amgaon Group
21. A21 Ampani Basin
22. A22 Anaipadi Formation
23. A23 Anceps Beds
24. A24 Andaman-Nicobar Islands Stratigraphy
25. A25 Andola Subgroup
26. A26 Angartoli Formation
27. A27 Anisoceras Beds
28. A28 Ankaleshwar Formation
29. A29 Annapurna Gneiss
30. A30 Aqtash Formation
31. A31 Aravalli Craton
32. A32 Aravalli Supergroup
33. A33 Archipelago Group
34. A34 Arenaceous Facies
35. A35 Ariyalur Group
36. A36 Arkasani Granite
37. A37 Arlikatti Dolomite
38. A38 Arsikere Granite
39. A39 Askot Gneiss
40. A40 Atal Volcanics
41. A41 Athgarh Beds
42. A42 Athleta Beds
43. A43 Avadi Formation
44. B1 Bababudan Group
45. B2 Babaguru Formation
46. B3 Babia Stage
47. B4 Badami Group
48. B5 Badhaura Formation
49. B6 Badhura Formation
50. B7 Bagalkot Group
51. B8 Bagh Beds/ Formation/ Group
52. B9 Bagra Formation
53. B10 Bahreli Formation
54. B11 Baijnath Gneiss
55. B12 Baila Formation

56. B13 Bailadala Group
57. B14 Baisakhi Formation
58. B15 Balcha Dhura Formation
59. B16 Balda Ganite
60. B17 Bandah Formation
61. B18 Bandal Gneiss
62. B19 Bansa-Chandiya Formation
63. B20 Bansi-Kakara Formation
64. B21 Bap Boulder Bed / Formation
65. B22 Baragolai Formation
66. B23 Barail Group
67. B24 Barakar Formation
68. B25 Baratang Group/ Formation
69. B26 Baripada Beds
70. B27 Barmer Formation
71. B28 Barotiya Group
72. B29 Barren Measures
73. B30 Barren Sandstones
74. B31 Basal Muschelkalk
75. B32 Basal Sandstone
76. B33 Basantgarh Group
77. B34 Bastar Craton
78. B35 Batal Formation
79. B36 Befliaz Formation
80. B37 Bengpal Group
81. B38 Berach Granite
82. B39 Bermoti Series
83. B40 Berwali Series
84. B41 Betul Supracrustal Belt/ Group
85. B42 Bhadasar Formation
86. B43 Bhaderah Formation
87. B44 Bhander Group
88. B45 Bhilwara Supergroup
89. B46 Bhim (Bheem) Group
90. B47 Bhima Basin/Group
91. B48 Bhimaram Formation
92. B49 Bhimdasa Formation
93. B50 Bhopalpatnam Granulites/ Belt
94. B51 Bhuana Formation
95. B52 Bhuban Formation
96 B53 Bhuj Beds
97. B54 Bhuj Member
98. B55 Bhumbli Member
99. B56 Bichom Formation
100. B57 Bijawar Basin/Group
101. B58 Bijori Formation
102. B59 Bikona Formation
103. B60 Bilara/Hansera Group/ Formation
104. B61 Birmania Basin/Formation
105. B62 Blaini Boulder Bed/ Group/ Formation
106. B63 Bokabil Formation
107. B64 Boldamgiri Formation
108. B65 Bombay Basin
109. B66 Bomdila Group
110. B67 Bonai Granite
111. B68 Borghat Formation
112. B69 Boulder Conglomerate Formation, Siwalik Group
113. B70 Boulder Conglomerate Formation, Trichinopoly
114. B71 Boulder Slate Formation
115. B72 Broach (Bharuch) Formation
116. B73 Budavada Sandstone
117. B74 Bundelkhand Craton
118. B75 Burtsa Formation
119. C1 Calcareous Sandstone
120. C2 Calcareous Shale
121. C3 Cambay Basin
122. C4 Cambay Shale Formation
123. C5 Cambrian of India
124. C6 Carboniferous of India
125. C7 Carboniferous of Kashmir
126. C8 Cauvery Basin

127. C9 Cave Deposits of Kurnool
128. C10 Cenozoic Era
129. C11 Chakradharpur Granite
130. C12 Chamba Formation
131. C13 Champaner Group
132. C14 Champion Gneiss
133. C15 Chanda Limestone Formation
134. C16 Chari Formation
135. C17 Charnokites
136. C18 Chaya Formation
137. C19 Chedrapal Outlier
138. C20 Cherakur Formation
139. C21 Cherla Formation
140. C22 Cherra Formation
141. C23 Chharat Limestone Formation
142. C24 Chhattisgarh Basin
143. C25 Chhattisgarh Supergroup
144. C26 Chhojan Member
145. C27 Chhongtash Formation
146. C28 Chhotanagpur Gneiss/ Gneissic Complex (CGC)
147. C29 Chidamu Beds
148. C30 Chikiala Formation
149. C31 Chikkim Formation
150. C32 Chikshellikeri Limestone
151. C33 Chilka Lake
152. C34 Chilpi Group
153. C35 Chinji Formation
154. C36 Chitradurga Granite
155. C37 Chitradurga Group
156. C38 Chitradurga Shear Zone
157. C39 Chitravati Group
158. C40 Chocolate Series
159. C41 Chomule Formation
160. C42 Chorgali Group
161. C43 Closepet Granite
162. C44 Coal Shale Facies
163. C45 Coastal Gondwana Formations
164. C46 Conglomerate Bed, Narmada valley
165. C47 Coral Limestone
166. C48 Coralline Limestone
167. C49 Coromandal Formation
168. C50 Cratons
169. C51 Cretaceous of Andaman and Nicobar Islands
170. C52 Cretaceous of Jammu & Kashmir
171. C53 Cretaceous of Pondicherry
172. C54 Cretaceous of Trichinopoly
173. C55 Cretaceous of Virdhachalam
174. C56 Cuddalore Sandstone Formation
175. C57 Cuddapah Basin
176. C58 Cuddapah Supergroup
177. D1 Dagshai Formation
178. D2 Dailing Group
179. D3 Dalbuing Formation
180. D4 Dalma Volcanic Group
181. D5 Dalmiapuram Formation
182. D6 Dalu Formation
183. D7 Damuda Group
184. D8 Daonella Limestone
185. D9 Daonella Shales
186. D10 Darik Member
187. D11 Debari Group
188. D12 Deccan Traps
189. D13 Dedza Formation
190. D14 Degana Granite
191. D15 Delhi Supergroup
192. D16 Delwara Group
193. D17 Denwa Formation
194. D18 Deoban Group
195. D19 Deoban Window
196. D20 Deola and Chirakhan Marl
197. D21 Dergaon Formation

198. D22 Devgarh Group
199. D23 Devonian of India
200. D24 Dhanjori Group
201. D25 Dharagad Group
202. D26 Dharamgarh Gneiss
203. D27 Dharangadhara Formation
204. D28 Dharangadhara Group
205. D29 Dharmaram Formation
206. D30 Dharwar Craton
207. D31 Dharwar Supergroup
208. D32 Dhekiajuli Formation
209. D33 Dhok Pathan Formation
210. D34 Dhosa Oolite
211. D35 Dhudamaniya Formation
212. D36 Diadematus Zone
213. D37 Dihing Formation
214. D38 Dilburi Formation
215. D39 Dilpur Formation
216. D40 Dinod Formation
217. D41 Dirang Group
218. D42 Disama Formation
219. D43 Disang Formation/Group
220. D44 Dogra Slates
221. D45 Dolam Sandstone Formation
222. D46 Doli Sandstone Formation
223. D47 Dongargarh Granite
224. D48 Dras Formation
225. D49 Dras Volcanics
226. D50 Drumgam Formation
227. D51 Dubrajpur Formation
228. D52 Dudatoli Formation
229. D53 Dupitila Formation
230. D54 Dwarka Formation
231. D55 Dykes of Malani Igneous Suite
232. E1 Eastern Ghats Mobile Belt
233. E2 Enchencheruvu Dolomite Formation
234. E3 Eocene of Andaman-Nicobar Islands
235. E4 Eocene of Jammu
236. E5 Eocene of Kashmir
237. E6 Eocene of NE India
238. E7 Eocene of Rajahmundry, Andhra Pradesh
239. E8 Eocene of Rajasthan
240. E9 Erinpura Granite Suite
241. E10 Exotic Blocks of Malla Johar
242. F1 Fatehgarh Formation
243. F2 Fenestella Shale
244. F3 Ferruginous Oolitic Formation
245. F4 Frog Beds
246. F5 Fulra Limestone Formation
247. G1 Gaj Formation
248. G2 Gajansar Beds
249. G3 Gamir Formation
250. G4 Ganeshpur Formation
251. G5 Gangamopteris Beds
252. G6 Gangapur Formation
253. G7 Ganmachidam Formation
254. G8 Garbyang Formation
255. G9 Garhwal Group
256. G10 Gechang Formation
257. G11 Geku Formation
258. G12 Ghuneri Member
259. G13 Giambal Gneiss
260. G14 Girujan Clays/Formation
261. G15 Giumal Formation
262. G16 Giumal Sandstone Formation
263. G17 Godavari Supergroup
264. G18 Godhara Granite
265. G19 Gogunda Group
266. G20 Golapalli Formation
267. G21 Golden Oolite

268. G22 Gondwana of Extra Peninsula

269. G23 Gondwana Supergroup

(a) Lower Division

(b) Upper Division

270. G24 Goru Formation

271. G25 Gorubathan Formation

272. G26 Gorur Gneiss

273. G27 Great Boundary Fault

274. G28 Grey Beds

275. G29 Grey Shale Formation

276. G30 Guitar Formation

277. G31 Gund Formation

278. G32 Gundul Formation

279. G33 Gungri Formation

280. G34 Gurnaghat Formation

281. G35 Gushal Formation

282. G36 Gwalior Basin/Group

283. H1 Habur Formation

284. H2 Haimanta Group

285. H3 Halkal Formation

286. H4 Halobia Beds

287. H5 Halorites Beds

288. H6 Hangrang Formation

289. H7 Hansera Group/Formation

290. H8 Hapatnar Group

291. H9 Harda Inlier

292. H10 Harudi Formation

293. H11 Harwal Shale

294. H12 Hedenstoemia Beds

295. H13 Higher Muschelkalk Member

296. H14 Hill Limestone Formation

297. H15 Himgiri Beds

298. H16 Himmatnagar/ Ahmadnagar Sandstone

299. H17 Hindoli Group

300. H18 Hirpur Formation

301. H19 Hosakatti Argillite

302. H20 Hoti Member

303. H21 Hundri Formation

304. H22 Hydrocarbons producing basins of India

305. i 1 Ichnology

306. i 2 Indo-Gangatic Plain

307. i 3 Indravati Basin

308. i 4 Indus Formation

309. i 5 Infra-trappean beds

1. Andhra Pradesh

(a) Marepalli

(b) Rajahmundry

2. Gujarat

(a) Rahioli

3. Madhya Pradesh

(a) Jabalpur

4. Maharashtra

(a) Dongargaon

(b) Pisdura

(c) Umrer

5. Krishna-Godavari Basin

(a) Chintalapalli-A

(b) Elamanchilli-A

(c) Modi-A

(d) Narsapur well-I

(e) Palakollu

310. i 6 Inglis Formation

311. i 7 Inolgiri Member

312. i 8 Inter-trappean beds

1. Andhra Pradesh

(a) Asifabad

(b) Rajahmundry

(c) Naskal

(d) Rangapur

2. Gujarat

(a) Anjar

(b) Bamanbor

(c) Kora

(d) Lakshmipur

3. Karnataka
(a) Biligi
(b) Gurmatkal
(c) Upparhatti
(d) Yellur
4. Madhya Pradesh
(a) Kisalpuri
(b) Mohgaon-Kalan
(c) Padwar/Ranipur
(d) Mandla
(e) Sagar
5. Maharashtra
(a) Bombay (Mumbai)
(b) Nagpur
6. Rajasthan
(a) Mamoni
7. Uttar Pradesh
(a) Lalitpur
8. Krishna-Godavari Basin
(a) Chintapalli
(b) Elamanchili
(c) Modi-A
(d) Narsapur Well-I
(e) Palakollu
(f) Razole
9. Rajmahal Basin
10. Chimalgutta
11. Malangutta

313. J1 Jabalpur Formation
314. J2 Jagannath or Dungoaposi Volcanics
315. J3 Jagat Group
316. J4 Jagdalpur Formation
317. J5 Jaintia Group
318. J6 Jaisalmer Formation
319. J7 Jakaram Arkose Formation
320. J8 Jalore Granite
321. J9 Jambusar Formation
322. J10 Jaunsar Group
323. J11 Jenam Formation
324. J12 Jhagadia Formation
325. J13 Jhamarkotra Formation
326. J14 Jhangu Formation
327. J15 Jharol Group
328. J16 Jodhpur Group/Formation
329. J17 Jogira Formation
330. J18 Jurassic of Kashmir
331. J19 Jurassic of Kutch
332. J20 Jurassic of Painkhanda
333. J21 Jutogh Group
334. J22 Juvavites Beds
335. K1 Kaga Formation
336. K2 Kailas (Kargil) conglomerate
337. K3 Kaimur Group
338. K4 Kakara Formation
339. K5 Kakdi Stage
340. K6 Kaladgi Basin
341. K7 Kalamund Formation
342. K8 Kalapani Limestone
343. K9 Kalhal Formation
344. K10 Kalkel Formation
345. K11 Kallakkudi Limestone Formation
346. K12 Kallamedu Formation
347. K13 Kallankurichi Formation
348. K14 Kamakhyanagar Quartzite
349. K15 Kamlial Formation
350. K16 Kamthi Formation
351. K17 Kanawar Group
352. K18 Kanchanjhanga Gneiss
353. K19 Kand/Kond Formation
354. K20 Kanger Formation
355. K21 Kanksawati Series
356. K22 Kanker Granitoids
357. K23 Kanthkot Sandstone
358. K24 Kaptipada granite
359. K25 Kapurdi Formation

360. K26 Karai Shale Formation
361. K27 Karal
362. K28 Karewa Group
363. K29 Kargil Conglomerate
364. K30 Karharbari Formation
365. K31 Karihul Formation
366. K32 Karikal Beds
367. K33 Karruchal Formation
368. K34 Kasauli Formation
369. K35 Katageri Limestone
370. K36 Katamdevarahalli Limestone
371. K37 Katesar Member
372. K38 Katrol Formation
373. K39 Kekri Formation
374. K40 Kerur Arenite
375. K41 Khairagarh Group
376. K42 Khalsi Formation
377. K43 Khari Nadi Formation
378. K44 Khari Series
379. K45 Khariar Basin
380. K46 Khasil Group
381. K47 Kherapara Formation
382. K48 Khetri Basin
383. K49 Khreuh Formation
384. K50 Khuiala Formation
385. K51 Khunamuh Formation
386. K52 Kimberlites
387. K53 Kiogad Formation
388. K54 Kioto Formation/Group
389. K55 Kioto Member
390. K56 Kirthar Formation
391. K57 Kohlan Group
392. K58 Koira Group
393. K59 Kolar Group
394. K60 Kondagaon Granulites/Belt
395. K61 Kopili Formation
396. K62 Kota Formation
397. K63 Krishna-Godavari Basin
398. K64 Krol Belt/Formation/Group
399. K65 Kuda Member
400. K66 Kuar Bet Beds
401. K67 Kuilapal Granite
402. K68 Kukti Formation
403. K69 Kulakkalnattam Formation
404. K70 Kuling Shale/Formation/Group
405. K71 Kullu Group
406. K72 Kundargi Quartzite
407. K73 Kunzun La Formation
(a) Spiti area
(b) Zanskar area, J & K
408. K74 Kurat Basalt
409. K75 Kurnool Group
410. K76 Kuti Shales
411. L1 Labenspurren
412. L2 Laisong Formation
413. L3 Lakadong Limestone Member
414. L4 Lakadong Sandstone Member
415. L5 Lakanavaram Shale Formation
416. L6 Laki Formation
417. L7 Lalab Formation
418. L8 Lamayuru Formation
419. L9 Lameta Formation
420. L10 Lamprophyre
421. L11 Langarin Member
422. L12 Langpar Formation
423. L13 Lapthal Member
424. L14 Largi Group
425. L15 Largi Window
426. L16 Lassaria Formation
427. L17 Laterite
428. L18 Lathi Formation
429. L19 Lidder Group
430. L20 Lignite Deposits
431. L21 Lilang Group

432. L22 Lilang Supergroup
433. L23 Limestone and Shale Member
434. L24 Lipak Formation
435. L25 Lochambal Beds
436. L26 Lokapur Subgroup
437. L27 Lolab Formation
438. L28 Long Formation
439. L29 Lower Katrol
440. L30 Lower Limestone Member
441. L31 Lower Muschelkalk
442. L32 Lumla Formation
443. L33 Lunavada Group
444. M1 Macchal Formation
445. M2 Macrocephalus Beds
446. M3 Madh Series
447. M4 Mahadek Formation
448. M5 Mahadeo Formation
449. M6 Mahadeva Formation
450. M7 Mahakoshal Supracrustal Belt/Group
451. M8 Mahakut Breccia
452. M9 Main Boundary Fault
453. M10 Main Central Thrust
454. M11 Malabar Supergroup
455. M12 Malangtoli (Nuakot) Volcanics
456. M13 Malani Granite
457. M14 Malani Igneous Suite
458. M15 Malanjkhand Granite
459. M16 Maleri Formation
460. M17 Malla Johar Supergroup
461. M18 Mallampalli Group
462. M19 Mamal Formation
463. M20 Mancharial Quartzite Formation
464. M21 Mandi Volcanics
465. M22 Manendragarh Marine Bed
466. M23 Mangalwar Complex
467. M24 Mangli Beds
468. M25 Maniyara Fort Formation
469. M26 Manjir Formation
470. M27 Mar Formation
471. M28 Maravattur Clay Formation
472. M29 Margo Formation
473. M30 Marh Formation
474. M31 Marwar Craton
475. M32 Marwar Supergroup
476. M33 Maspur Basalt Formation
477. M34 Mastoli Formation
478. M35 Matanomadh Formation
479. M36 Mathaji ka Dungar Formation
480. M37 Mayurbhanj Granite
481. M38 Mayyanad Formation
482. M39 Meekoceras zone
483. M40 Megalodon Limestone/Para Stage
484. M41 Meghalaya Group
485. M42 Mesozoic Era
486. M43 Mesozoic succession of Kutch, Gujarat
487. M44 Mewar Gneiss
488. M45 Middle Katrol
489. M46 Mikin Formation
490. M47 Mikir Hills Trap
491. M48 Milan Group
492. M49 Miliolite / Porbandar Stone
493. M50 Miliolite/Miliolite Limestone
494. M51 Mineralised Supracrustal Belts
495. M52 Miocene rocks of Karnataka
496. M53 Miocene rocks of Maharashtra
497. M54 Miri Formation
498. M55 Mobile Belts
499. M56 Mochia/Zawar Formation
500. M57 Monotis Shales
501. M58 Moran Group

502. M59 Morar Formation
503. M60 Motur Formation
504. M61 Muddapur Dolomite
505. M62 Mulug Group
506. M63 Murree Formation
507. M64 Muschelkalk Lower
508. M65 Muschelkalk Upper
509. M66 Muth Quartzite/ Formation
510. N1 Naga Parbat Gneiss
511. N2 Nagaur Basin
512. N3 Nagaur Group/Formation
513. N4 Nagri Formation
514. N5 Nagum Formation
515. N6 Nallamalai Group
516. N7 Nambi Formation
517. N8 Namsang Formation
518. N9 Nandgaon Group
519. N10 Nareda Formation
520. N11 Narha Beds
521. N12 Nari Formation
522. N13 Narmada Formation
523. N14 Narpuh Sandstone Member
524. N15 Neil West Coast Formation
525. N16 Nepheline Syenite of Kishangarh
526. N17 Nerinea Beds
527. N18 Newania Carbonatite
528. N19 Newer Dolerite
529. N20 Neyveli Lignite Field
530. N21 Nimar Sandstone
531. N22 Nimoloksa Group
532. N23 Niniyur Group/Formation
533. N24 Niralkeri Breccia
534. N25 Nishatbagh Bed
535. N26 Niti Limestone Member
536. N27 Niti Member
537. N28 Nodular Limestone, Bagh
538. N29 Nodular Limestone, Spiti
539. N30 Nunuluka Formation
540. O1 Offshore Basins of India
541. O2 Older Metamorphic Group (OMG)
542. O3 Older Metamorphic Group of NE India
543. O4 Older Metamorphic Tonalite Gneiss (OMTG)
544. O5 Ongarbira Volcanics
545. O6 Ophiceras Zone
546. O7 Ordovician-Silurian rocks
547. O8 Otoceras Zone
548. O9 Ottakovil Formation
549. P1 Pachcham Basal Beds
550. P2 Pachcham Coral Bed
551. P3 Pachcham Formation
552. P4 Pachcham Shell Limestone
553. P5 Pachmarhi Formation
554. P6 Pairi Group
555. P7 Pala Lahara Gneiss
556. P8 Palana Formation
557. P9 Palaram Shale Formation
558. P10 Paleozoic Era
559. P11 Paleozoics of Kashmir
560. P12 Pali Formation
561. P13 Panchet Formation
562. P14 Panchet Group
563. P15 Pandyan Mobile Belt
564. P16 Panjal Volcanics
565. P17 Papaghni Group
566. P18 Papaghni Sub-basin
567. P19 Par Formation
568. P20 Para Stage
569. P21 Parahio Series
570. P22 Parh Formation
571. P23 Pariwar Formation
572. P24 Parsoi Formation
573. P25 Parsora Formation
574. P26 Pashkyum Formation
575. P27 Passage Beds

576. P28 Pattipalli Quartzite Formation
577. P29 Penganga Group
578. P30 Peninsular Gneiss
579. P31 Permian rocks
580. P32 Phulad Lineament
581. P33 Phulad Ophialite Suite
582. P34 Pindobal Formation
583. P35 Pinjor Formation
584. P36 Piram Beds
585. P37 Pleistocene sediments
    (a) Kurnool, Ghod, Manjara, Pravara and Godavari areas
    (b) Central Narbada (Upper Group), Palmar
    (c) Central Narbada (Lower Group)
    (d) Lower Karewa, Pinjor, Upper Shiwalik
586. P38 Po Formation
587. P39 Pokran Boulder Bed
588. P40 Porbandar Stone
589. P41 Porlob Group
590. P42 Port Blair Formation
591. P43 Port -Meadow Formation
592. P44 Potla Formation
593. P45 Prang Member/Formation
594. P46 Pranhita Sandstone Formation
595. P47 Pranhita-Godavari Basin
596. P48 Pranhita-Godavari Valley
597. P49 Precambrian of Himalaya
598. P50 Productus Limestone
599. P51 Productus Shale
600. P52 Proterozoic Sedimentary Basins
601. P53 Punagarh Group
602. P54 Purana Basins
603. P55 Putnur Limestone Formation
604. Q1 Qazil langer Formation
605. Q2 Quartzite Series
606. Q3 Quatenary sediments of Himachal Pradesh
607. Q4 Quaternary sediments of Andhra Pradesh
608. Q5 Quaternary sediments of Central India
609. Q6 Quaternary sediments of Gujarat Mainland
610. Q7 Quaternary sediments of Kali Valley, Kumaon, Uttarakhand
611. Q8 Quaternary sediments of Maharashtra
612. Q9 Quaternary sediments of Tamil Nadu
613. Q10 Quaternary sub-Era
614. R1 Rabanpalli Formation
615. R2 Raghvapuram Formation
616. R3 Raialo Group
617. R4 Rajahmundry Formation
618. R5 Rajgarh Group
619. R6 Rajmahal Basin
620. R7 Rajmahal Formation
621. R8 Rajmahal Inter-trappeans
622. R9 Rajmahal Traps
623. R10 Rama Formation
624. R11 Ramagiri Formation
625. R12 Ramania Stage
626. R13 Rambakot Member
627. R14 Ramban Formation
628. R15 Ramnad Beds
629. R16 Rampur Group
630. R17 Ramsu Formation
631. R18 Randha Formation
632. R19 Rangamal Formation
633. R20 Rangapahar Formation
634. R21 Rangapani Formation
635. R22 Rangrik Formation

636. R23 Raniganj Formation
637. R24 Ranikot Formation
638. R25 Ranipet Formation
639. R26 Ranthambore Group
640. R27 Ras Marble
641. R28 Ratanpur Member
642. R29 Rawali Bagar Group
643. R30 Recent
644. R31 Rehmanni Beds
645. R32 Renji Formation
646. R33 Rewa Group
647. R34 Rewak Formation
648. R35 Rikhabdeo/Rishabhdeo ultramafics
649. R36 Rimkhim Member
650. R37 Rishkobal Formation
651. R38 Rohtang Gneiss
652. R39 Rongtong Formation
653. R40 Round Formation
654. S1 Sabari Basin/Group
655. S2 Sakoli Group
656. S3 Salkhala Group (with Tanwal Formation)
657. S4 Salooni Formation
658. S5 Sancha Malla Group
659. S6 Sandhan Formation
660. S7 Sandmata Complex
661. S8 Sangeluma Group
662. S9 Sanglung Formation/Group
663. S10 Sanu Formation
664. S11 Sargur Group
665. S12 Sarnu Formation
666. S13 Saser Brangsa/Shaksgam Formation
667. S14 Satnala Shale Formation
668. S15 Satpura Mobile Belt
669. S16 Satyavedu Formation
670. S17 Sauna Volcanics
671. S18 Saundatti Quartzite
672. S19 Sausar Supracrustal Belt/Group
673. S20 Sawai Bay Formation
674. S21 Se La Group
675. S22 Sedam Subgroup
676. S23 Semri Group
677. S24 Serpentine Group
678. S25 Seti Khola Formation
679. S26 Shahabad Limestone
680. S27 Shal Shal Formation
681. S28 Shali Group
682. S29 Shali Window
683. S30 Shell Limestone
684. S31 Shiala Formation
685. S32 Shillong Group
686. S33 Shimoga Schist Belt or Super Belt
687. S34 Shumahal Formation
688. S35 Shumar Formation
689. S36 Siju Formation
690. S37 Sillakudi Formation
691. S38 Silurian of Kashmir
692. S39 Silurian of Kumaon, Uttarakhand
693. S40 Silurian of Spiti, Himachal Pradesh
694. S41 Simikeri Subgroup
695. S42 Simla Slate/Group
696. S43 Simlipal Basin/Group
697. S44 Sincha Formation
698. S45 Sindreth Group
699. S46 Singhbhum Craton
700. S47 Sirmur Group
701. S48 Sirohi Group
702. S49 Sivaganga Beds
703. S50 Sivaganga Formation
704. S51 Siwalik Group
(a) Lower Division
(b) Middle Division
(c) Upper Division

705. S52 Siwana Granite or Siwana Ring Complex
706. S53 Sleemanabad Formation
707. S54 Solda Formation
708. S55 Somadevara Quartzite Formation
709. S56 Somanpalli Group
710. S57 Sonakhan Group
711. S58 SONATA (Son-Narmada-Tapti) Lineament
712. S59 Sonrai Basin
713. S60 Sonrai Formation
714. S61 Spiti Formation/Shale
715. S62 Spiti Member
716. S63 Sriperumbudur Formation
717. S64 Strait Formation
718. S65 Subathu Formation/Group
719. S66 Suket Shales
720. S67 Sukma Group
721. S68 Sulcacutus Bed
722. S69 Sullavai Group
723. S70 Sundergarh Group
724. S71 Supra Panchet Formation
725. S72 Surajdeval Formation
726. S73 Surma Group
727. S74 Suru Gneiss
728. S75 Suwana Formation
729. S76 Sylhet Limestone Formation
730. S77 Sylhet Sub-group/Formation
731. S78 Sylhet Trap
732. S79 Syringothyris Limestone
733. T1 Tagling Limestone
734. T2 Takalpalli Arkose Formation
735. T3 Takche/Manchap Formation
736. T4 Takli Formation
737. T5 Tal Group/Formation
738. T6 Talchir Formation
739. T7 Tamba Kurkus Group
740. T8 Tanawal/Conglomerate/ Formation
741. T9 Tandi Group
742. T10 Tannaki/Tanawal/Langera Conglomerate
743. T11 Tarakeshwar Formation
744. T12 Tatrot Formation
745. T13 Tectonic framework of India
746. T14 Terani plant beds
747. T15 Teri sands
748. T16 Teris
749. T17 Tertiary
750. T18 Tertiary of Gujarat
751. T19 Tertiary of Kashmir
752. T20 Tethyan Basin
753. T21 Thabo Beds
754. T22 Than Formation
755. T23 Thango Formation
756. T24 Therria Formation
757. T25 Thimpu Group
758. T26 Tikak Parbat Formation
759. T27 Tikki Formation
760. T28 Tindra Formation
761. T29 Tipam Formation/Group
762. T30 Tipam Sandstone Formation
763. T31 Tippapuram Shale Formation
764. T32 Tiratgarh Formation
765. T33 Tirodi Gneiss
766. T34 Tirupati Formation
767. T35 Trace Fossils
768. T36 Traumatocrinus Beds
769. T37 Triassic of Byans
770. T38 Triassic of Jammu and Kashmir
771. T39 Triassic of north Kumaon and Garhwal
772. T40 Triassic of NW Himalaya
773. T41 Triassic of Painkhanda

774. T42 Triassic of Spiti
775. T43 Trichinopoly (Tiruchirapalli) area
776. T44 Trichinopoly Group
777. T45 Trigonia Beds
778. T46 Trigonoarca Beds
779. T47 Tropites Beds
780. T48 Tso Morari Gneiss
781. T49 Tura Formation
782. T50 Tusham/Tosham Ring Complex, Haryana
783. U1 Udaipur Group
784. U2 Ukra Beds
785. U3 Ukra Member
786. U4 Um Sohryngkew Formation
787. U5 Umaria Marine Bed
788. U6 Umia ammonite Bed
789. U7 Umia Beds
790. U8 Umia Formation
791. U9 Umia Member
792. U10 Umia Plant Beds
793. U11 Umia Stage
794. U12 Umlatdoh Limestone Member
795. U13 Upper Assam Basin
796. U14 Upper Katrol Sandstone
797. U15 Upper Katrol Shales
798. U16 Upper Muschelkalk
799. U17 Upper Sandstone Formation
800. U18 Usur Group
800. U19 Uttapalli Conglomerate Formation
802. U20 Uttarlai Formation
803. U21 Uttatur Group
804. V1 Vagadkol Formation
805. V2 Vaikom Formation
806. V3 Vaikrita Group
807. V4 Valudavur Beds
808. V5 Variegated Formation
809. V6 Vejendla-Vemavaram Formation
810. V7 Vemavaram Formation
811. V8 Vembanad Formation
812. V9 Venkatapur Sandstone Formation
813. V10 Venkatapuram Group
814. V11 Vindhyan Basin
815. V12 Vindhyan Supergroup
816. V13 Vinjhan Shale Formation
817. V14 Vinjhan Stage
818. W1 Wadhwan Formation
819. W2 Waior Stage
820. W3 Wangtu Gneiss
821. W4 Warkalli Group
822. W5 Wazur Formation
823. W6 Wumuh Formation
824. W7 Wuyan Formation
825. Y1 Yadahalli Argillite
826. Y2 Yamne Formation
827. Y3 Yargatti Argillite
828. Y4 Yashwantanagar Formation
829. Y5 Yerrapalli Formation
830. Y6 Yinkiang Group
831. Z1 Zamia Beds
832. Z2 Zamia Shales
833. Z3 Zawar/Mochia Formation
834. Z4 Zewan Formation

■■■■

III

DESCRIPTION OF FORMATIONS

(Total 834 Formations described)

A 1 Abor Volcanic Formation: A Lower Permian age formation of Arunachal Pradesh, best exposed in road section around Dosing, along Pangin Geku-Yinkiong road. The volcanic rocks comprise basalt, andesite, acid tuffs, lapillis and agglomerates. The volcanic underlie the Geku Formation of Paleocene age of Yinkiang Group. See Yinkiang Group and Table 22.

A 2 Abujhmar Basin: A Puran Basin of Paleoproterozoic age covering about 3000 sq.km. area in the Bastar Craton and housing the Abujhmar Group. See Abujhmar Group and Tables 4 and 32.

A 3 Abujhmar Group: A Proterozoic age Group occupying the Abujhmar Basin and overlying the Bailadala and Nandgaon Groups in the Kotri-Dongarhgarh Orogen of Bastar Craton. The Group consists of lower Gundul Formation, which consists basically of conglomerates, and Maspur Basalt Formation, consisting mainly of gabbro/dolerite sills and dykes. See Abujhmar Basin and Table 4.

A 4 Agglomeratic Slate: In Kashmir, Laddakh and Eastern Korakoram in the State of J & K, this is Carboniferous to Permian age formation, overlying the Fenestella Shale. In the revised classification Agglomeratic Slate is equivalent to Pindobal Formation. See Table No. 11

A 5 Ahmadnagar Sandstone: See Himmatnagar Sandstone.

A 6 Aida Stage: An Oligocene-Miocene age stage of Kutch, Gujarat, which lies above the Waior Stage of Bermoti Series, of which it is probably the uppermost stage. Aida Stage is also encountered sub-surface in Lakhpat I Well, Kutch, Gujarat, consisting of fossiliferous variegated siltstone. See Tables 20 and 38.

A 7 Aishmuqam Formation: See Table 11

A 8 Ajabgarh Group: A Group of Delhi Supergroup, lying over the Alwar Group in the Bayana-Lalsot and Alwar Basins of the North Delhi Fold Belt. In the Bayana-Lalsot Basin, the Ajabgarh Group consists of carbonaceous phyllite in the lower part and ferruginous quartzite in the upper part. In the Alwar Basin the Ajabgarh Group at the base consists of carbonates and phosphorites interlayered with basic volcanic, followed by alternation of quartzite, carbon phyllite, mica schist and marble.

A 9 Ajmer Formation: An associate group of South Delhi Fold Belt of Dehli Supergroup, occurring north of Bithur-Pisagan, near Sambhar salt-lake, consisting of quartzite and arkose-pelite-greywacke suite that overlies Anasagar migmatites, Ajmer, Rajasthan.

A 10 Akli Formation: An Eocene age formation of Barmer Basin in western Rajasthan consisting of fossiliferous sandstone, bentonite, bentonitic clay, siliceous earth, ferruginous shale and phosphatic sandstone with gypsum and pyrite.

A 11 Alalor Formation: An Upper Triassic formation of Lilang Group, Spiti, consisting of coral limestone bed. See Table 13

A 12 Albaka Pamanuru (Chelimala) Formation: See Godavari Supergroup .

A 13 Albitite: It is a 300 km. long zone of albitite, an uncommon metamorphic rock found intruded into the North Dehli Fold Belt of the Delhi Supergroup and basement gneisses and are associated with ultramafic rocks and are known for uraninite-fluorite-magnetite-chalcopyrite-molybdenite mineralization.

A 14 Alluvium, Newer: The light coloured and calcium poor alluvium, predominantly of Holocene (Recent) age, occurring in the Indo-Gangatic Plain along with the Older Alluvium and wind-blown deposits e.g. loess and desert sand. Newer Alluvium also called 'Khadar' contain fossil parts of elephants, horses, crocodiles, etc. In delta region of Ganga and Brahmputra rivers the Newer Alluvium consists of clay, sand and marl with layers of peat, lignite, etc. Varanasi Alluvium which is of Middle to Late Pleistocene age is also included here in the Newer Alluvium.

A 15 Alluvium, Older: A comparatively dark coloured alluvium of Middle to Upper Pleistocene age of Indo-Gangatic Plain occurring in small patches and narrow bands forming slightly elevated terraces along flood level in the Ganga valley region of U.P. and Bihar. The alluvium of Peninsular rivers e.g. Narmada, Tapti (Tapi), Mahanadi, Godavari, Krishna, Penner, Cavery, etc. are also included in the older Alluvium. The Older Alluvium also called 'Bhangar', comprises multicyclic sequences of sand, silt and clay rich in calcium carbonate nodules and concretions called 'kankar' contain fossil parts of many mammals e.g. elephas, eqqus, rhinoceros, etc.

A 16 Almora Gneiss: A Precambrian gneiss of Lesser Himalaya in Himachal Pradesh. See Precambrian of Himalaya.

A 17 Alwar Basin Granites: The Alwar Basin of the North Dehli Fold Belt of the Delhi Supergroup is intruded by several granitic bodies like Bairath Granite, Ajitgarh Granite, Barodia Granite and Harsoria Granite.

A 18 Alwar Group: It lies over Raialo Group in the Bayana-Lalsot Basin and Alwar Basin of North Dehli Fold Belt of Delhi Supergroup. In Bayana-Lalsot Basin the Group consists of conglomerate, orthoquartite, felspathic quartzite and arkose along with subordinate shale, while in Alwar Basin the Group consists of quartzite and conglomerate.

A 19 Ambalapuzha Formation: The uppermost formation of Warkalli Group. See Table 25.

A 20 Amgaon Group: Oldest Group of Archean age of Kotri-Dongargarh Orogen of Bastar Craton. In the east of Sakoli Belt Amgaon Group enclaves consist of quartzites, calc-silicate rocks, marbles, garnet-staurolite-kyanite-sillimanite mica schist, cordierite-anthophyllite rocks, amphibolites, chromiferous ultra-mafics and subordinate banded iron formation. Most enclaves are similar to that of Sukma Group, but for the staurolite-kyanite-sillimanite mica schists. See Table 4.

A 21 Ampani Basin: It is one of the Proterozoic Sedimentary Basins (Purana Basins) of Bastar Craton of Meso-Neoproterozoic age and contains 280m thick sequence of sandstone and shale. The Basin is associated with Keskal, Singanpur and Chedrapal outliers. See Table 32.

A 22 Anaipadi Formation: See Trichinopoly area.

A 23 Anceps Beds: See Table 15 Mesozoic succession of Kutch, Gujarat.

A 24 Andaman-Nicobar Islands Stratigraphy: Consists of Upper Cretaceous Paleo-gene Group and Archipelago Group. See Table 21 for details of formations and lithology.

A 25 Andola Subgroup: See Bhima Basin/Group.

A 26 Angartoli Formation: A Miocene age formation, equivalent to Bokabil Form-ation, occurring in the Garo Hill area of Assam-Arakan Basin in NE India. See Table 23

A 27 Anisoceras Beds: See Cretaceous of Pondicherry.

A 28 Ankaleshwar Formation: A Middle Eocene age subsurface formation of Gujarat Mainland. See Table 20 Cenozoic Formations of Gujarat.

A 29 Annapurna Gneiss: A Precambrian gneiss of Lesser Himalaya in Uttarakhand. See Precambrian of Himalaya.

A 30 Aqtash Formation: A Permian age Formation of Karakoram Range in Hunza and Laddakh area of Jammu and Kashmir.

A 31 Aravalli Craton: The Aravalli Craton occupies an area of over one lakh sq.km., covering entire Rajasthan and parts of Gujarat, Madhya Pradesh, Dehli and Haryana. Eastern part of Aravalli Craton is occupied by Mewar Craton and western part by Marwar Craton. Aravalli Craton is unique in the sense that it has formations of complete geological time-scale, right from Precambrian to Cenozoic and houses 2 full Supergroups, the Aravalli and Delhi and part of Vindhyan Supergoup. The main Precambrian formations of Aravalli Craton are shown in Table 6.

A 32 Aravalli Supergroup: In the Aravalli Craton, overlying the Mangalwar Complex and Sandamata Complex of the Bhilwara Supergroup and Mineralised Supracrustal Belts of Archean Era, lies the Aravalli Supergroup of Palaeoproterozoic age. The Supegroup has following sequence:

Igneous Intrusions of Slumbar (Duda) Granite - (Youngest)

Champaner Group

Lunavada Group

Rikabdeo ultamafics

Jharol Group

Udaipur Group

Debari Group

Delwara Group (oldest)

A 33 Archipelago Group: A group comprising of Miocene and younger rocks in the Andaman and Nicobar Islands.

A 34 Arenaceous Facies: Represents the lower part of Barail Group of Oligocene age in the Upper Assam Valley. See Table 23

A 35 Ariyalur Group: See Trichinopoly area and Table 18.

A 36 Arkasani Granite: A granite- granophyre rock of the younger Granitoids of the Singhbhum Craton, consisting of massive core of monzo-granite to syeno-granite and schistose rim.

A 37 Arlikatti Dolomite: See Kaladgi Basin and Table 37.

A 38 Arsikere Granite: A granite-granophyre rock of the younger granitoids of the Chaibasa Formation of the Singhbhum Craton, consisting of massive core of monzo-granite to syeno-granite and schistose rim.

A 39 Askot Gneiss: A Precambrian gneiss of Lesser Himalaya in Himachal Pradesh. See Precambrian of Himalaya.

A 40 Atal Volcanics: A Precambrian volcanic associated with Dharagad Group. See Precambrian of Himalaya.

A 41 Athgarh Beds: Lower Cretaceous beds of the Gondwana Supergroup, occurring at the head of the Mahanadi delta, near Cuttack on the east coast and consisting of conglomerates, grits, carbonaceous shale etc.

A 42 Athleta Beds: See Table 15.

A 43 Avadi Formation: A cretaceous formation of Palar Basin occurring on the east coast of Andhra Pradesh and containing fossils of ammonites, belemnites, bivalves and foraminifera.

B 1 Bababudan Group: The oldest group of the Dharwar supergroup in the western Dharwar Craton. From bottom to top the Group consists of Kalasapura, Allanpura, Santaveri, and Mulaingiri Formations, consisting of metabasalts, gabbros, ultamafic schists, phyllites, quartzites, Banded Iron Formation, etc.

B 2 Babaguru Formation: A Lower Miocene age formation overlying the Eocene age Dinod Formation of Gujarat Mainland, consisting of 200-300 mts. Of highly ferruginous sandstone and agate conglomerate. See Table 20.

B 3 Babia Stage: An Middle Eocene age, upper stage of Berwali Series, Kutch, Gujarat, also encountered in the Lakhpat-I well in Kutch, Gujarat. See Tables 20 and 38.

B 4 Badami Group: See Kaldgi Basin

B 5 Badhaura Formation: A Permian age marine formation of Bikaner-Nagaur Basin in western Rajasthan. See Table 16.

B 6 Badhura Formation: Same as Badhaura Formation.

B 7 Bagalkot Group: See Kaladgi Basin

B 8 Bagh Beds/Formation/Group: A Cretaceous age formation of Narmada Basin occurring as detached outcrops along the north bank of Narmada river from Barwani in Khargon district of Madhya Pradesh, reaching to Chota Udaipur in Vadodra district and Rajpipli in Bharuch district of Gujarat. Bagh Formation consist of richly fossiliferous sandstone, nodular limestone, coralline limestone and marl containing fossils of Ammonoidea/Ammonites, Lamellibranchia, Gastropods and Echinoidea. Succession from bottom to top is Nimar Sandstone, Nodular limestone, Deola and Chirakhan marl and coralline Limestone. Lameta Formation which is also of Cretaceous age is just younger to the Bagh Formation.

B 9 Bagra Formation: A Gondwana Supergroup formation of Upper Triassic age of Satpura Basin consisting of conglomerate and overlying the Denwa Formation.

B 10 Bahreli Formation: A Permian age, Lower Gondwana formation of Arunachal Pradesh, overlying the Bichom Formation, consisting of fedspathic sandstone and shale with coal beds and plant fossils.

B 11 Baijnath Gneiss: A Precambrian gneiss of Lesser Himalaya in Himachal Pradesh. See Precambrian of Himalaya.

B 12 Baila Formation: A Precambrian formation of Lesser Himalaya in Himachal Pradesh. See Precambrian of Himalaya.

B 13 Bailadala Group: Is of Proterozoic age and overlies the Bengal Group in the Bastar Craton. The Group predominantly consists of banded iron formation and felspathic quartzite, conglomerate, ultramafic rocks, chlorite and sericite phyllites and polymicts. See and Table 4.

B 14 Baisakhi Formation: A Jurassic age formation of Jaisalmer Basin, Western Rajasthan, consisting of fossiliferous gypseous shale, argillaceous sandstone and interformational conglomerate.

B 15 Balcha Dhura Formation: See Malla Johar Supergroup and Table 17.

B 16 Balda Ganite: A granitic intrusion of Marwar Craton of Western Rajasthan, occurring as phacolithic dykes and apophyses consisting of non-porphyritic leucogranite and hosting tungsten mineralization.

B 17 Bandah Formation: A richly fossiliferous Eocene age formation of Jaisalmer Basin in Western Rajasthan, overlying the Khuiala Formation and consisting of bioclastic limestone. See Table 16.

B 18 Bandal Gneiss: A Precambrian gneiss of Lesser Himalaya in Himachal Pradesh. See Precambrian of Himalaya.

B 19 Bansa-Chandiya Formation: A Gondwana Supergroup formation of Lower Cretaceous age in the Rewa area of Son-Mahandi Basin, consisting of white clays, silty sandstone, silt and sandstone.

B 20 Bansi-Kakara Formation: See Table 22.

B 21 Bap Boulder Bed /Formation: Occurs at village Bap, near Phalodi in Jodhpur district in western Rajasthan. The boulders, pebbles and cobbles, some of them have ice-scratched like marks consist of sandstone and limestone. Earlier the formation was considered of Permo-Carboniferous age, bur recent studies have established a post-Eocene age, possibly Pleistocnne age. Moreover the boulders etc. do not occur as a bed, but as loose pell-mell spread over surface without any matrix and are closely associated with Quaternary sediments. See Table 16.

B 22 Baragolai Formation: The middle formation of Barail Group of Oligocene age in Naga Hill Fold Belt. See Table 23.

B 23 Barail Group: The Group comprises of Oligocene age rocks of NE region, which includes Mikir Hills, Dhansari valley, Upper Assam, etc. areas. See Tables 22 and 23.

B 24 Barakar Formation: A Lower Permian age formation of Gondwana Supergroup consisting of sandstone, shale, coal deposits, etc. Barakars of Pranhita-Godavari Valley are richly fossiliferous. See Table 12.

B 25 Baratang Group/ Formation: See Cretaceous of Andaman and Nicobar Islands and Table 21.

B 26 Baripada Beds: Miocene age beds of Baripada area in Orissa, consisting of clay and sand.

B 27 Barmer Formation: A Paleocene-Eocene age formation of Western Rajasthan, consisting of red and brown sandstone. See Table 16.

B 28 Barotiya Group: Middle Group of the Sendra Basin of South Fold Belt of the Delhi Supergroup, consisting of Barr Conglomerate, mica schist, impure marble, subarkose and bimodal volcanic.

B 29 Barren Measures: A Middle Permian age formation of Gondwana Supergroup, overlying the Barakar Formation. The formation is devoid of coal seams and consists of ferruginous sandstone, micaceous siltstone and ferruginous shale or the iron stone shale. Lower Pali and Motur are its equivalent formations. See Table 12.

B 30 Barren Sandstones: A Jurassic age bed of Katrol Formation of Kutch, Gujarat. See Table 15.

B 31 Basal Muschelkalk: A Middle Triassic bed of Lilang Group, Spiti, consisting of 1 m of fossiliferous shaly limestone. See Table 13.

B 32 Basal Sandstone: Lowermost formation of Jaintia Group of Eocene age in the Upper Assam Valley. See Table 23.

B 33 Basantgarh Group: The oldest Group of the Sendra Basin, South Fold Belt, Delhi Supergroup, consisting of metabasalt, calc-schist, metapelite, quartzite, marble, mica schist, etc.

B 34 Bastar Craton: Situated almostly in the central part of Peninsula, from SE of Nagpur upto Eastern Ghats. See Table 4.

B 35 Batal Formation: A middle to late Cambrian Formation of the Zanskar-Kashmir Tethyan sub-basin forming the basal unit of the Haimanta Group and consisting of micaceous quartzite, flaggy quartzite, pyritous carbonaceous phyllite, gritstone and some limestone. See Table 11. Batal Formation of Terminal Proterozoic age however also appears in the IV and III cycle of Higher Himalaya of Himachal Pradesh in the Precambrian stratigraphy of Himalaya.

B 36 Befliaz Formation: A Precambrian formation of Lesser Himalayas of Jammu and Kashmir. See Precambrian of Himalaya.

B 37 Bengpal Group: Is of Neoarchean age and belongs to Kotr-Dongargarh Orogen of Bastar Craton and extends from Bailadila to Tulsi Dungar Hills and consists of quartz sericite schists and sericite quartzites, Amygdular meta-basalts, chloritoid-chlorite schist, andalusite phyllite and schists and andalusite-garnet quartzires. See Table 4.

B 38 Berach Granite: Covers a large area west of Chittorgarh in Rajasthan in the Aravalli Craton and belongs to the Hindoli Group of the Bhilwara Supergroup. It consists of an older foliated granodiorite to adamellite and younger grey granite to leucogranite and is intruded with pegmatite and aplite.

B 39 Bermoti Series: An Oligocene age Series of Kutch, Gujarat, consisting of lower Ramania Stage and Upper Waior Stage. Ramania Stage consists of greenish grey marl and argillaceous limestone with a basal boulder clayey marl bed. Waior Stage consists of a basal oolitic band followed by fossiliferous marl. Aida Stage which is of Oligocene-Miocene age probably forms the top part of Bermoti Series. See Tables 20 and 38.

B 40 Berwali Series: An Eocene age formation of Kutch, Gujarat, consisting of lower Kakdi Stage and upper Babia Stage. Kakadi Stage cossnsists of black shales with lignite bands, bitumen and pyritous shale. Babia Stage consists of fossiliferous limestone. See Tables 20 and 38.

B 41 Betul Supracrustal Belt/Group: Occurs in the Central Indian Tectonic Zone of the Satpura Mobile Belt above the Tan Shear Zone and below the Balrampur Fault. The Belt from bottom to top consists of Golighat Group, Amla Gneiss, Bargaon Group, Bhopali Group, Padhar Mafic Intrusion and Younger Granite.

B 42 Bhadasar Formation: A Jurassic (Tithonian) formation of Jaisalmer Basin in Western Rajasthan, consisting of fossiliferous oolitic grits, calcareous sandstone and conglomerate. See Table 16

B 43 Bhaderah Formation: A Precambrian Formation of Higher Himalaya of Himachal Pradesh. See Precambrian of Himalaya.

B 44 Bhander Group: The Uppermost Group of the Vindhyan Supergroup. For details see Table 9.

B 45 Bhilwara Supergroup: The Archean age Bhilwara Supergroup in the Aravalli Craton forms basement for the Aravalli Supergroup and includes part of earlier called Banded Gneissic Complex (BGC). The Supergroup broadly consists of 3 groups which from bottom to top are Hindoli Group, Mangalwar Complex (along with isolated mineralized Belts) and Sandmata Complex. The mutual stratigraphic relations between these units and with the Aravalli – Delhi Orogene are controversial. See Table 6.

B 46 Bhim (Bheem) group: Upper group of the Bhim Basin of South Delhi Fold Belt of Delhi Super Group, consisting of pelitic to semi-pelitic schists, calc-silicate rocks and wollastonite bearing impure marble. See Table 8.

B 47 Bhima Basin/Group: It is a Proterozoic (Meso-Neoproterozoic) Sedimentary Basin (Purana Basin) of Dharwar Craton, occupied by Bhima Group. The Bhima Group consists of lower Sedam Subgroup and Upper Andola Sub-group. The Sedam Subgroup consists of Rabanpalli Formation and Shahabad Limestone, while Andola Subgroup consists of lower Halkal Formation, middle Katamdevarahalli Limestone and upper Harwal Shale. See Table 32

B 48 Bhimaram Formation: See Pranhita-Godavari Valley.

B 49 Bhimdasa Formation: A Precambrian formation of Lesser Himalaya of Jammu and Kashmir. See Precambrian of Himalaya.

B 50 Bhopalpatnam Granulites/Belt: Belongs to Bastar Craton and is about 300 km long and 30-40 km wide and is of Archean age and consists of grey gneisses and migmatites of TTG gneiss composition. Quartzites are rich in sillimanite, calc-silicate rocks contain wollastonite, scapolite and humite. See Table 4.

B 51 Bhuana Formation: A Triassic age formation which is not exposed but occurs only subsurface in the Jaisalmer Basin of Western Rajasthan. See Table 16.

B 52 Bhuban Formation: Lower formation of Surma Group/Series of Miocene age occurring in the Khasi and Jaintia Hills, Naga Hills Fold Belt and Assam-Arakan Basin in NE India. See Table 23.

B 53 Bhuj Beds: See Table 15.

B 54 Bhuj Member: See Umia Stage

B 55 Bhumbli Member: Middle member of Miocene age of Gaj Formation of Saurashtra, Gujarat, consisting of conglomerate. See Table 20.

B 56 Bichom Formation: A formation of Bomdila Group in Arunachal Pradesh. The formation is also named as Bomte, Garu and Rilu Formations in different areas. The Bicom formation is divided into Rilu, Bomte and Sensa Members.

B 57 Bijawar Basin/Group: It is one of the Purana Basins of Paleoproterozoic age of the Bundelkhand Craton, occurring in Madhya Pradesh around Bijawar town. The basin houses the Bijawar Group. Stratigraphy of Bijawar Group has been shown in Table 34. See Tables 32 and 34.

B 58 Bijori Formation: A Gondwana Supergroup formation of Upper Permian age and equivalent to Raniganj Formation in the Satpura Basin, consisting predominantly of clays in the eastern part and sandstone in the western pare. See Table 12.

B 59 Bikona Formation: A pliocene age formation equivalent to Dupitila Formation of Garo Hills of Assam-Arakan Basin of NE India. See Table 23.

B 60 Bilara/Hansera Group/Formation: A formation of Nagaur Basin of Marwar Supergroup of Proterozoic to Cambrian age lying above Jodhpur Formation/ Group and below Nagaur Formation/Group. Hansera and Bilara Groups are homotaxial facies variant of each other and both consist of non-phosphatic stromatolites. Bilara Group consists of limestone and dolomite with intercalations of shale, siltstone and pebble beds . See Table 16.

B 61 Birmania Basin/Formation: The Marwar Supergroup was deposited in two basins, larger the Nagaur Basin and smaller the Birmania Basin. Birmania Basin is of Proterozoic to Cambrian age and consists of lower sandstone formation and upper carbonate formation. The sandstone formation is correlated with the Jodhpur Sandstone of Nagaur Basin and consists of shale, sandstone and grit bed. The overlying carbonate contains basal needle shale, cherty and dolomitic limestone, phosphorite and sandstone. See Table 16.

B 62 Blaini Boulder Bed/Group/Formation: Basically a boulder bed formation with some slates and limestone in the Lesser Himalaya of Shimla area of Himachal Pradesh. Earlier Blaini Boulder Bed were believed to be of Carboniferous age, equivalent to the Gondwana Talchir Boulder Bed, but now it is established to be of Vendian-Early Cambrian age. See Precambrian of Himalaya.

B 63 Bokabil Formation: The upper formation of Surma Group/Series of Miocene age occurring in Khasi and Jaintia Hills, Mikir Hills and Dhansari Valley, upper Assam Valley, Naga Hills Fold Belt, Surma Valley and Tripura Fold Belt areas of Assam – Arakan Basin of NE India. See Tables 22 and 23.

B 64 Boldamgiri Formation: A Miocene age formation equivalent to Bhuban Formation, occurring in the Garo Hills area of Assam-Arakan Basin of NE India. See Table 23.

B 65 Bombay Basin: See Off Shore Basins and Table 29.

B 66 Bomdila Group: A Precambrian (Paleoproterozoic) formation of Lesser Himalaya in Arunachal Pradesh. See Precambrian of Himalaya.

B 67 Bonai Granite: A rock type of the Singhbhum Craton consisting predominantly of porphyritic granite with small patches of mymatitic trondhjomite and older suracrystal enclaves. See Table 5.

B 68 Borghat Formation: A Cretaceous age formation of Borghat-Narphu section near Shillong Plateau, Khasi and Jaintia Hills in NE India, consisting of sandstone/boulder conglomerate and lying below the Mahadek Formation or it may be a part of Mahadek Formation. See Table 23.

B 69 Boulder Conglomerate Formation, Siwalik Group: Uppermost Formation of Pleistocene age of Siwalik Group. See Table 28.

B 70 Boulder Conglomerate Formation, Trichinopoly : See Trichinopoly area and Table 18.

B 71 Boulder Slate Formation: These are Permian age fossiliferous rocks of Lesser Himalaya of Doggada and adjoining area of Garhwal in Uttarakhand and are similar to the Agglomeratic Slate of Kashmir.

B 72 Broach (Bharuch) Formation: A Pliocene age subsuraface fossiliferous formation of Gujarat Mainland, consisting of claystone and sandstone. See Table 20.

B 73 Budavada Sandstone: A cretaceous age formation of Ongole area in Andhra Pradesh, may be equivalent to Rajmahal Traps/sediments of Gondwana Supergroup.

B 74 Bundelkhand Craton: It is smallest of all the Cratons with an area of 26000 sq.kms. lying around Chhatarpur, Jhansi, Gwalior, etc. in Madhya Pradesh, consisting of Archean age Bundelkhand Granite with supracrystal enclaves traversed by numerous quartz reefs and include swarms of mafic dykes. The Bundelkhand Granite consists of hornblende granodiorite, pophyritic biotite granite, coarse pophyritic granite and leucogranite.

B 75 Burtsa Formation: An Upper Triassic to Cretaceous age fossilferous formation of Karakoram Range in Hunza and Laddakh areas of Jammu and Kashmir State.

C 1 Caslcareous Sandstone: A Permian age sandstone bed of Spiti area, Himachal Pradesh.

C 2 Calcareous Shales: Lower bed of Cretaceous age of Umia Formation, Kutch, Gujarat. See Table 15.

C 3 Cambay Basin: See Offshore Basins and Table 29.

C 4 Cambay Shale Formation: A Lower Eocene age sub-surface formation of Gujarat Mainland. See Table 20.

C 5 Cambrian of India: Cambrian rocks in India occur in Jammu and Kashmir (Kashmir Valley and Zanskar sector), in Himachal Pradesh (Lahaul-Spiti-Kinnaur sub-basin) and in Uttarakhand (Gayang Formation, Kumaon).

C 6 Carboniferous of India: Carboniferous rocks occur in Kashmir sub-basin. In Himachal Pradesh and Uttarakhand Carboniferous rocks form the Kanawar Group. Talchir Formation of Gondwana Supergroup is of Carboniferous to Lower Permian age.

C 7 Carboniferous of Kashmir: In Kashmir sub-basin Carboniferous formations from bottom to top are Tannaki Conglomerate, Syringothyris Limestone, Fenestella Shale and Agglomeratic Slate. Recently Syringothyris Limestone along with Passage Beds has been renamed as Aishmuqam Formation, Fenestella Shale as Ganeshpur Formation and Agglomeratic Slate as Pindobal Formation. Aishmuqam, Ganeshpur and Pindobal Formations together have been put under Lidder Group. In Karakoram area of Kashmir Saser Brangsa/ Shaksgam Formation are of Late Carboniferou age.

C 8 Cauvery Basin: See Offshore Basins and Table 29.

C 9 Cave Deposits of Kurnool: In Kurnool district of Andhra Pradesh, caves known as Billa Sargam caves in the Kurnool Group Limestone of Neoproterozoic age, contain red marls deposits of Upper Pleistocene age on the cave floors with bones of apes, mammals, birds, reptiles and carnivorous animals. In Kashmir Valley Harwan cave has Holocene age deposits.

C 10 Cenozoic Era: It is latest of all the eras and belongs to Phanerozoic Eon. See Table 19 for subdivisions and age.

C 11 Chakradharpur Granite: A gneissic granite of Singhbhum Craton, belonging to Younger Ganitoids and containing amphibolites, talc schist and arenaceous sediments. See Table 5.

C 12 Chamba Formation: A Precambrian formation of Mesoproterozoic age in the Higher Himalya in Himachal Pradesh. See Precambrian of Himalaya.

C 13 Champaner Group: Youngest group of Aravalli Supergroup, consisting of sub-greywakes, siliceous phyllites, pelitic schists, quartzites and petromict conglomerate. Bands of dolomitic limestone and manganiferous phyllite near Shivrajpur are economically important. See Table 6.

C 14 Champion Gneiss: Uppermost member of Kolar Group of Dharwar Supergroup of the Eastern Dharwar Craton. See Table 3.

C 15 Chanda Limestone Formation: A Formation of Proterozoic Godavari Supergroup. See the Supergroup and Table 30.

C 16 Chari Formation: A Jurassic Formation of Kutch, Gujarat. See Tables 15 and 12.

C 17 Charnokites: Proterozoic age Charnockites overlie the Archean age Dharwar Supergroup in the Dharwar Craton. The rock was named by geologist T.H. Holland in 1900 after Job Charnock- the founder of Kolkata city, whose tomb is built of this rock. Charnockite is not a single rock, but a suite or series of non-foliated metamorphic rocks of different types. Some of the charnockites are felsic and rich in quartz and microcline; others are mafic and full of pyroxene and olivine. There are also intermediate varieties corresponding mineralogically to norite, quartz-norite and diorites. Hypersthene is a commonly occurring mineral in most of the charnokites. Charnokites are found in south India in the Nilgiri Hills, the Shevaroys, the Biligirirangan Hills and part of the Western Ghats, extending southwards to Cape Comorin. See Table 2.

C 18 Chaya Formation: A Holocene age formation of Saurashtra coast, Gujarat, consisting of semi-consolidated to consolidated shell limestone, coral reefs and oyster beds. See Table 20.

C 19 Chedrapal outlier: See Sabri (Sukma) Basin/Group.

C 20 Cherakur Formation: See Indravati Basin.

C 21 Cherla Formation: A Proterozoic Godavari Supergroup formation. See the Supergroup and Table 30.

C 22 Cherra Formation: See Therria Formation

C 23 Chharat Limestone Formation: A Eocene age formation of Potwar-Jammu area of Jammu and Kashmir State. See Table 22.

C 24 Chhattisgarh Basin: It is one of the largest Puraana Basins of Meso-Neoproterozoic age, lying in the Bastar Craton. The Basin consists of conglomerate, shale, sandstone, limestone and dolomite. See Table 32.

C 25 Chhattisgarh Supergroup: The Supergroup lies in the Chhattisgarh Basin of Bastar Craton and is of Proterozoic age and is divisible into 3 Groups, which from bottom to top are Singhora, Chandrapur and Raipur Groups. See Tables 32 & 33.

C 26 Chhojan Member: Middle Member of Kiogad Formation of Jurassic age in the Niti area of Painkhanda, Uttarakhand, consisting of oolitic ferruginous arenaceous limestone.

C 27 Chhongtash Formation: An early Permian age fossiliferous formation of Karakoram Range in Hunza and Laddakh area of Jammu and Kashmir.

C 28 Chhotanagpur Gneiss/Gneissic Complex (CGC): It forms one of the 4 sectors of the Satpura Mobile Belt. The Chhotanagpur Gneiss consists of following lithotectonic units- Tamar-Poropahar Shear Zone, Gumla-Purulia Granulite Belt, Ranchi Supracrystal Belt, Hazaribagh-Dumka Granulite Belt, Hesathu-Belbathan Belt, Bihar Mica Belt and Munger Belt.

C 29 Chidamu Beds: See Spiti /Shale Formation.

C 30 Chikiala Formation: Uppermost Gondwana Supergroup formation of Lower Cretaceous age occurring with the Gangapur Formation in the Godavari Valley area and consisting of Unaossiliferous conglomerate, conglomeratic sandstone and ferruginous sandstone. See Pranhita-Godavari Valley and Table 12.

C 31 Chikkim Formation: A fossiliferous Upper Cretaceous formation of Spiti area, consisting of Lower Limestone Member and Upper Shale Member. The Lower Limestone Member consists of limestone, dolomitic limestone with alternation of shale and marl. The Upper Shale Member consists of grey silty shale, shaly limestone, marl and fine siltstone bands.

C 32 Chikshellikeri Limestone: See Kaladgi Basin and Table 37.

C 33 Chilka Lake: A gradually sinking shallow lake- a lagoon of Pleistocene age on the east coast near Puri in Orissa, separated from sea (Bay of Bengal) by sand barriers or spits. Above hightide level there are deposits of estuarine fossil shells of *Arca granosa, Meretrix costa, Sunetta seripta, Potamides angulatus* and *Cuma disjuncta*.

C 34 Chilpi Group: Is of Proterozoic age and is the youngest Group of Bastar Craton and consists of conglomerate at the base, but the main rocks of the Group are chloritic and sericitic phyllites and banded charts with associated crystalline limestone. See Table 4.

C 35 Chinji Formation: A Miocene age formation of Siwalik Group. See Siwalik Group and Table 28.

C 36 Chitradurga Granite: One of the youngest granites of Western Dharwar Craton. See Table 2.

C 37 Chitradurga Group: The upper group of the Dharwar Supergroup in the Western Dharwar Craton, consisting of lower the Vanivilas sub-Group and upper the Ranibennus sub-Group. The Vanivilas sub-Group consists of conglomerates, metabasalts and phyllites, followed by mafic-felsic volcanics with Banded Iron Formation (BIF) and phyllite that include manganese and iron formations, stromatolitic carbonate rocks, cherts, pelites, quartzites and conglomerates. The Ranibennur sub-Group consists of greywackes, Banded Iron Formation, conglomerates and volcanics. See Tables 2 and 3.

C 38 Chitradurga Shear Zone: The Shear Zone separates the Western Dharwar Craton from the Eastern Dharwar Craton. The contact between western and Eastern Dharwar Cratons is not sharp but transitional. See Table 2.

C 39 Chitravati Group: This is the second oldest group of the Cuddapah Supergroup. The Group consists of 3 formations which from bottom to top are Pulivendla Quartzite, consisting of quartzite and conglomerate, the Tadpatri Formation consisting of shale, felsic tuff, basic salts and flows, chert, stromatolitic dolomite and quartzite and Gandikota Quartzite consists of shale, quartzite and glauconitic quartzite. See Table 7.

C 40 Chocolate Series: A Triassic age formation of northern Kumaon in Byans, Uttarakhand.

C 41 Chomule Formation: An Upper Triassic formation of Spiti, Himachal Pradesh, consisting of Halobia Beds and Daonella Limestone Bed. See Triassic of Spiti and Lilang Supergroup and Tables 13 and 14.

C 42 Chorgali Group: A Devonian age formation of Lidder Valley, Kashmir. See Table 11.

C 43 Closepet Granite: It is one of the younger granites of the Eastern Dharwar Craton and forms a 400 km. long and 20-30 km wide plutonic belt which is divisible into southern and northern parts, while based on the mode of emplacement it is divisible into 3 depth zones, which are Root Zone, Transfer Zone and Intrusive Zone. See Table 2.

C 44 Coal Shale Facies: Represents the upper part of Barail Group of Oligocene age in the Upper Assam Valley. See Table 23.

C 45 Coastal Gondwana Formations: These occur in the Ramnathpuram area of Tamil Nadu, along with the Sivaganga Beds and contain plant and marine fossils and are of Lr. Cretaceous (Aptian) age.

C 46 Conglomerate Bed, Narmada valley: This is a fossiliferous conglomerate bed of Middle Pleistocene age, in the Narmada valley in Central India, in which fossilized skull of Narmada Homo erectus was found along with other fossils.

C 47 Coral Limestone: A 30 m limestone bed of Upper Triassic age of Lilang Group in Spiti, Himachal Pradesh.

C 48 Coralline Limestone: A Cretaceous (Senonian) age bed of Bagh Beds. See Bagh Beds.

C 49 Coromandal Formation: See Quaternary deposits of Tamil Nadu.

C 50 Cratons: See Tectonic framework of India

C 51 Cretaceous of Andaman and Nicobar Islands: The Late Cretaceous to Paleocene sediments and intrusive cover almostly the entire Island and comprises of Porlob Group along with associated igneous rocks forming the Serpentine Group, overlain by Baratang Group. The porlob Group consists of porcellaneous limestones or marbles, cherts, jaspers and quartzites. Serpentine Group consists of gabbros, peridotites, dunites and serpentines. Baratang Group consists of marine grey shales interbedded with limestones and siltstones and is exposed in the Baratang Island.

C 52 Cretaceous of Jammu and Kashmir: The Cretaxeous rocks of Jammu and Kashmir are represented by Sangeluma Group in Laddakh and by Hundri

Formation in the Eastern Korakoram. See Sangeluma Froup and Hundri Formation

C 53 Cretaceous of Pondicherry: A small but richly fossiliferous exposure of Cretaceous rocks occurring around Valudavur near Pondicherry, consisting of limestones and shales is classified into 3 divisions, namely- 1. Valudavur or Anisoceras Beds, 2. Trigonoarca Beds and 3. Nerinea Beds. The first two divisions are correlated with the Ariyalur Group and the third with the Niniyur Group. See Table 18.

C 54 Cretaceous of Trichinopoly: The best developed and most fossiliferous rocks of India, occurring in the Trichinopoly area of Tamil Nadu, comprising from bottom to top of Uttatur, Trichinopoly and Ariyalur Groups. See Trichinopoly area and Table 18.

C 55 Cretaceous of Virdhachalam: A small outcrop of Cretaceous rocks of Ariyalur Group occur in Virdhachalam area of Trichinopoly District of Namil Nadu. See Trichinopoly area and Table 18.

C 56 Cuddalore Sandstone Formation: An Upper Miocene to Pliocene age formation occurring as disconnected outcrops at Vriddhachalam, Ariyalur, Thanjavur and Sivaganga in Tamil Nadu. The formation generally consists of gritty, pebbly to conglomerate, coarse to fine grained argillaceous sandstone, somewhat ferruginised and laterised towards top with intercalated clays. The formation contain extensive lignite beds, e.g. the Neyveli Lignite Deposit and huge petrified tree trunks near Tiruvakkarai which is declared as National Geological Monument.

C 57 Cuddapah Basin: A Meso-Neoproterozoic sedimentary Purana Basin of Dharwar Craton, occupied by Cuddapah Supergroup and Kurnool Group. See Cuddapah Supergroup and Kurnool Group and Tables 7 and 32.

C 58 Cuddapah Supergroup: The Supergroup is of Meso-Neoproterozoic age and lies over the Crystalline Basement of the Dharwar Craton. The Supergroup from botton to top consists of 3 Groups, namely Papaghni, Chitravati and Nallamalai. See each Group separately and Table 7.

D 1 Dagshai Formation: A formation of Late Eocene to Oligocene age of Sirmur Group, overlying the Subathu Formation in the Lesser Himalaya of Himachal Pradesh, comprising of lower Chimnun and upper Pabo Member. Chimnun Member consists of 700 mts. of red and purple claystone with subordinate siltstone, while the Pabo Member consists of 1200 mts. of altenating sandstone and clay. See Table 22.

D 2 Dailing Group: A Precambrian Group of Lesser Himalaya in Sikkim-Darjeeling area.

D 3 Dalbuing Formation: See Yinkiang Group.

D 4 Dalma Volcanic Group: A Group of Archean age Singhbhum Craton. The Group consists of a lower and an upper formations. The lower formation consists of carbon phyllite, tuff and quartzite interlayered with volcanic. The upper formation consists of high-Mg (picritic to komatiitic) volcanoclastics and komatiites along with co-magmatic intrusion. See Table 5.

D 5 Dalmiapuram Formation: A Lower Cretaceous formation of Uttatur Group of Trichinopoly area. The Formation is also called Kallakkudi Limestone Formation. See Table 18.

D 6 Dalu Formation: A formation equivalent to the upper part of the Dihing Formation of Pliocene to Pleistocene age in the Assam-Arakan Basin of NE India. See Table 23.

D 7 Damuda Group: A Group of Lower Gondwana Supergroup, comprising Karharbari, Barakar, Barren Measures, and Raniganj Formations in the Damodar Valley. See Table 12.

D 8 Daonella Limestone: A Middle Triassic Bed of Lilang Group of Spiti, Himachal Pradesh, consisting of 45 mts. of fossiliferous dark limestone, See Table 13.

D 9 Daonella Shales: A Middle Triassic bed of Lilana Group in Spiti, Himachal Pradesh, consisting of 50 mts. of fossiliferous black limestone, shaly limestone and shale. See Table 13.

D 10 Darik Member: Lower Member of Kherapara Formation of Oligocene age, equivalent to Barail Group in the Garo Hills area of NE India. See Table 23.

D 11 Debari Group: The second oldest group of the Aravalli Supergroup, overlying the Delwara Volcanics. The Group consists of petomict conglomerate (class of granite, gneiss, fuchsite, quartzite and mica schist), pebbly arkose and felspathic quartzite. The conglomerate extends from Ekling ji to Salumbar, near Udaipur. The Group includes Berwas and Jhamarkotra Formations which overlie the Petromict Conglomerate. Berwas Formation consists of calcareous quartzite, carbonaceous phyllite, dolomite, etc. Jhamarkotra Formation in addition to calcareous quartzite, carbonaceous phyllite, etc. include varieties of carbonate rocks. Grey dolomite is widely developed and includes stromatolitic phosphorite and phosphoratic dolomite. Jhamarkotra Formation near Udaipur is mined for phosphorite. See Table 6.

D 12 Deccan Traps: Popularly known Deccan Traps is actually Deccan Volcanic Province in geological literature and it is world's most distinguished Continental Flood Basalt Provinces. The Deccan Traps have a step-like topography and are divisible into Upper and Lower Divisions. Deccan Traps cover about 5 lacs sq.km in western and central India and have maximum thickness of about 3000 mts. The volcanic traps consist mostly of tholeitic basalts with some picrites, picritic basalts and alkaline basalts. The volcano-plutonic complexes consist of acid to alkaline intrusions and flows, gabbros and dolerites as well as lamprophyres and carbonatites. The Deccan Volcanics have erupted closer to Cretaceous-Tertiary boundary between 63 to 66 Ma, i.e. from Late Cretaceous to early Paleocene and might have caused extinction of Dinosaurs. A typical vertical stratigraphic coloumn through Deccan volcanic consists of 5 formations, which from bottom to top are Jawahar Formation, Igatpuri Formation, Thakurvadi Formation, Bhimashankar Formation and Khandala Formation. These formations consists of picrite, olivine phyric, Mafic phyric, Giant Plagioclase phyric, Plagioclasr Phyric, Amygdaloidal and Aphyric massive. Deccan Traps (Deccan Volcanic Province) is broadly divided into 4 sub-provinces, which are 1. Deccan Platrau (western, central and southeastern areas), 2. Malwa Plateau (areas around Sagar, Bhopal and Indore), 3. Mandla Lobe (easternmost outlier around Jabalpur, Seoni and Chhindwara) and 4. Saurashtra Plateau (bounded by Khambat graben in the east and western extension of Son-Narmada fault in the south). Sedimentary fossiliferous horizons found lying below the Deccan Traps are called Infra-trappeans which are found at Bagh, Lameta, Nagpur, Dharangadhra, Rajahmundry and Wardha. Fossiliferous sedimentary horizons found in between Trap layers are called Inter-trappeans. A few important Inter-trappeans are Anjar Inter-trappeans of Kutch, Gujarat which consists of 4 Inter-trappean beds within 7 Trap layers. Mumbai Inter-trappeans occur in Malabar, Cumballi and Worli Hills with famous Frog Beds. Inter-trappeans of Naskal, Pargi Taluka, Ranga Reddi District, Andhra Pradesh is noted for its mammalian fossils. Other Inter-trappeans are of Asifabad and Nagpur, Rajahmundry and Narasapur.

D 13 Dedza Formation: A Precambrian formation of Lesser Himalaya in Arunachal Pradesh. See Precambrian of Himalaya.

D 14 Degana Granite: A granite intrusion of the Marwar Craton, consisting of a topaz bearing Rapakivi granite and granite porphyry with tungsten mineralization and enriched in Rubidium (Rb), Lithium (Li) and Fluorine (F).

D 15 Delhi Supergroup: The Supergroup is of Mesoproterozoic age and overlies the Aravalli Supergroup in the Aravalli Craton and consists of 2 Fold Belts- older the South Delhi Fold Belt (SDFB) and younger the North Delhi Fold Belt (NDFB). See Tables 6 and 9.

D 16 Delwara Group: The oldest Group of the Aravalli Supergroup, best exposed near Nathdwara and continues over 300 km upto Pipalkhut and includes pockets of paleosol and consists of pebbly orthoquartzite overlain by Delwara volcanic consisting of amygdular basic flows intercalated with pyroclastic and epiclastic deposits including volcanic volcanic conglomerates. See Table 6.

D 17 Denwa Formation: A Gondwana Supergroup formation of Middle Triassic age in the Satpura Basin consisting of sandstones and clays. See Table 12.

D 18 Deoban Group: A Precambrian Group of Lesser Himalaya in Himachal Pradesh. See Precambrian of Himalaya.

D 19 Deoban Window: A structure in the second Precambrian Sedimentary Cycle of Lesser Himalaya in Himachal Pradesh, where the succession consists of lower Dharagarh Group and upper Deoban Group (earlier called Chakrata Series).

D 20 Deola and Chirakhan Marl: A Cretaceous (Cenomanian to Turonian) age beds of Bagh Beds Formation. See Bagh Beds Formation.

D 21 Dergaon Formation: A formation of Khalsi Group of Late Cretaceous age in the Mikir Hills and Dhansiri valley area of Assam-Arakan Basin in NE India.

D 22 Devgarh Group: A group of uncertain stratigraphic position closely associated with the South Delhi Fold Belt (SDFB) of the Delhi Supergroup consisting of quartzites and pelites and occurring at the periphery of the SDFB as a narrow band within basement gneiss to the east of Bhim Group.

D 23 Devonian of India: See Muth Quartzites/Formation.

D 24 Dhanjori Group: A Group of Singhbhum Craton, considered equivalent to Simlipal Group. The lower formation of the Group consists of pebbly phyllite with minor mafic-ultramafic intrusions, while the upper formation at the base consists of arkose, conglomerate and phyllite with volcanics at the top. See Table 5.

D 25 Dharagad Group: A Precambrian group of Lesser Himalaya in Himachal Pradesh with associated Atal Volcanics. See Precambrian of Himalaya.

D 26 Dharamgarh Gneiss: A Precambrian gneiss of Lesser Himalaya in Himachal Pradesh. See Precambian of Himalaya.

D 27 Dharangadhara Formation: See Dharangadhara Group.

D 28 Dharangadhara Group: Consists of Late Jurassic to Early Cretaceous fossiliferous sedimentary rocks of Surendranagar district, Saurashtra, Gujarat. The Group was also called as Kathiawar Beds, is divided into Dharangadhara Formation and Wadhwan Formation. The Group covers an area of about 500 sq.km. and consists of sandstone with lenses of conglomerate, shale and siltstone and has a total maximum thickness of about 600 m. The Group also includes Than, Surajdeval and Ranipet Formations.

D 29 Dharmaram Formation: See Maleri Formation and Pranhita-Godavari Valley.

D 30 Dharwar Craton: The Dharwar Craton is largest of all the cratons of Peninsular India occupying a major part of south India. Hyderabad, Chennai, Bangalore etc. cities lie within this craton. The Dharwar Craton is divisible into Western Dharwar Craton (WDC) and Eastern Dharwar Craton (EDC). The WDC and EDC are separated by Chitradurga Shear Zone. For statigraphic succession see Table 2.

D 31 Dharwar Supergroup: It is the largest unit of the Dharwar Craton and is Archean in age. In the Western Dharwar Craton the Supergroup is exposed in a large schist belt called Super Belt consisting of lower Bababudan Group and upper Chitradurga Group. In the Eastern Dharwar Craton the Supergroup consists of lower Yashwantanagar Formation and upper Kolar Group. See Tables 2 and 3.

D 32 Dhekiajuli Formation: A formation of Pliocene to Pleistocene age, equivalent to the Dihing Formation of the Upper Assam Valley in NE India. See Table 23.

D 33 Dhok Pathan Formation: See Siwalik Group and Table 28.

D 34 Dhosa Oolite: A bed of Chari Formation of Jurassic age in the Mesozoic Succession of Kutch. See Table 15.

D 34 Dharangadhara Group: Consists of Late Jurassic to Early Cretaceous fossiliferous sedimentary rocks of Surendranagar district, Saurashtra, Gujarat. The Group was also called as Kathiawar Beds, is divided into Dharangadhara Formation and Wadhwan Formation. The Group covers an area of about 500 sq.km. and consists of sandstone with lenses of conglomerate, shale and siltstone and has a total maximum thickness of about 600 m. The Group also includes Than, Surajdeval and Ranipet Formations.

D 35 Dhudamaniya Formation: The uppermost formation consisting of alternating sequence of Banded Iron Formation and phyllite of the Mahkoshal Group of Satpura Mobile Belt of Proterozoic age.

D 36 Diadematus Zone: A zone of Macrocephalus Beds of the Chari Formation of Jurassic age in the Mesozoic succession of Kutch. See Table 15.

D 37 Dihing Formation: A Pleistocene age formation of the Naga Hills Fold Belt, Surma Valley and Tripura Fold Belt areas of Assam-Arakan Basin of NE India. In Garo Hills Dihing Formation is represented by lower Rangapani and upper Dalu Formations, while in Mikir Hills and Dhansiri valley by Disama Formation. See Table 23.

D 38 Dilburi Formation: The uppermost formation of Tandi Group of Late Paleozoic age (probably Permian) of Himachal Pradesh.

D 39 Dilpur Formation: A Pleistocene formation of Karewa Group of Jammu and Kashmir. See Table 24.

D 40 Dinod Formation: An Eocene age formation of Gujarat Mainland, consisting of compact massive yellow fossiliferous limestone. See Table 20.

D 41 Dirang Group: A Precambrian Group of Lesser Himalaya in Arunachal Pradesh. See Precambrian of Himalaya

D 42 Disama Formation: A Pliocene to Pleistocene age formation (equivalent to Dihing Formation) of Mikir Hills and Dhansiri Valley of the Assam-Arakan Valley in NE India. See Table 23.

D 43 Disang Formation/Group: Paleocene to Eocene age rocks of NE India, equivalent to Jaintia Group, occurring in the southeastern part of Arunachal Pradesh in the Naga-Patkai Range. The formation contain impressions of dicotyledonous plant leaves and fossils of tiny lamellibranchs, gastropods and rare foraminifers. Formation also occurs in Surma Valley and Tripura Fold Belt. See Table 23.

D 44 Dogra Slates: A Cambrian Formation of Kashmir Tethyan sub-basin.

D 45 Dolam Sandstone Formation: A Proterozoic formation in the Albaka Belt of Usur Group of Godavari Supergroup. See the Supergroup and Table 30.

D 46 Doli Sandstone Formation: A Proterozoic formation in the Albaka Belt of Usur Group of Godavari Supergroup. See the Supergroup and Table 30.

D 47 Dongargarh Granite: A Proterozoic age Granite that intrudes Amgaon Gneiss and Nandgaon Groups of Kotri-Dongargarh Orogen in the Bastar Craton. See Table 4.

D 48 Dras Formation: See Sangeluma Group.

D 49 Dras Volcanics: See Sangeluma Group.

D 50 Drumgam Formation: A Jurassic age formation of Warwan valley, Doda district of Kashmir Himalaya, comprising of Member A (lower) consisting of fossiliferous quartzite, greywacke and slate and Member B (upper) consisting of fossiliferous sandy calcareous shale.

D 51 Dubrajpur Formation: A Gondwana Supergroup formation equivalent to the Spupra-Panchet formation of Upper Triassic age in the Rajmahal Basin in Jharkhand and West Bengal and consisting of feldspathic sandstone with some red and green shales. See Table 12.

D 52 Dudatoli Formation: A Precambrian formation of Higher Himalaya in Uttarakhand. See Precambrian of Himalaya.

D 53 Dupitila Formation: A Pliocene age formation of Khasi and Jaintia Hills, Surma Valley and Tripura Fold Belt in the Assam-Arakan Basin of NE India. Bikona, Rangapahar and Namsang are its equivalent formations in the same basin. See Table 23.

D 54 Dwarka Formation: A Miocene-Pliocene age formation of Okha area, northwestern Saurashtra, Gujarat, consisting of silty clays and arenaceous foraminiferal limestone. See Table 20.

D 55 Dykes of Malani Igneous Suite: A number of dykes consisting mostly of dolerite, some are microgranite, andesite, rhyolite porphery, etc. cut across rocks of Proterozoic Malani Igneous Suite. These dykes need to be distinguishe fom the Deccan Traps.

E 1 Eastern Ghats Mobile Belt: The Eastern Ghats Mobile Belt is one of the 3 Proterozoic Mobile Belts of the Peninsular India, the other two are the Pandyan Mobile Belt and the Satpura Mobile Belt. The Eastern Ghats Mobile Belt extends in NE-SW direction along the eastern coast for about 1000 kms. from Brahmani river in Orissa to Ongole in Andhra Pradesh. The Belt which is wide at the NE end tappers down at the SW end and has maximum thickness of 300 kms. From NE to SW the belt comes in contact with Singhbhum, Bastar and Eastern Dharwar Cratons. Different authors have given their own classification for the Eastern Ghats Mobile Belt. Ramakrishnan (1998) divided the Belt into Western Charnockite Zone, Western Khondalite Zone, Central Migmatite Zone and Eastern Khondalite Zone. Rickers (2001) divided the Belt into 3 groups of Domains- Domains IA, IB & IV; Domain II and Domain III. Dobmeier and Rath divided the Belt into 4 Provinces- Krishna Province (including Nellore-Khammam schist belt and Ongole Domain), Jeypore Province, Rangali Province and Eastern Ghats Province.

E 2 Enchencheruvu Dolomite Formation: A formation of Mulug Group of Pakhal/ Mallampalli Belt of the Godavari Supergroup. See the Supergroup and Table 30.

E 3 Eocene of Andaman-Nicobar Islands: Comprises of Baratang Formation consisting of shale and grewacke of late Createous to Eocene age and Port Blair Formation consisting of conglomerate, grit and greywacke of Middle Eocene to Oligocene age. See Table 21.

E 4 Eocene of Jammu: Eocene rocks of Jammu region in J & K State forms a 3 to 5 km. wide belt, which along with underlying Permo-Carboniferous rocks form a autochthonous belt between Murree and Panjal Thrusts. The lower part of the belt consists of 100 to 150 mts. of thin bedded cherty limestone with a few fossils of nummulites and gastropods. The upper part consists of pyritous coaly and ferruginous shales with thin carbonaceous beds. There are also a series of outcrops of Eocene rocks above the 'Great Limestone' inliers of Rajouri-Riasi areas consisting of laterite, pyritous-carbonaceous shales and Nummulitic Limestone. The shales overlying them contain workable coal seams.

E 5 Eocene of Kashmir: Eocene rocks of Kashmir consists of jaspery quartzites, shales and fossiliferous limestones and unaossiliferous red, purple and green shales with sandstone, overlying the Triassic limestone in Uri area forming recumbent anticline. In Kargil the Eocene limestone is highly shattered and extends for about 300 kms. along the upper Indus Valley. Laddakh Molasse Group consisting of conglomerates, sandstones and shales is divided into Kargil, Tharumsa and Pashkyum Formations.

E 6 Eocene of NE India: Eocene rocks of NE India belong to the Jaintia Group and equivalent formations. See these and Table 23.

E 7 Eocene of Rajahmundry, Andhra Pradesh: Consists of certain inter-trappean beds having algal fossils of Acicularia, Neomaria, Chara, etc.

E 8 Eocene of Rajasthan: Eocene of Rajasthan include Kapurdi, Akli, etc. formations of Barmer Basin, Banda and Khuiala Formations of Jaisalmer Basin and Jogira Formation of Bikaner-Nagaur Basin. See Table 16.

E 9 Erinpura Granite Suite: A granite intrusion in the Mewar Craton of the Aravalli craton. The intrusion is about 325 long and 3 to 15 km wide and consists of granite which is commonly porphyritic but locally gneissic and includes Mt. Abu and Sewariya Granites.

E 10 Exotic Blocks of Malla Johar: These are blocks of white to pinkish limestone with fossils of Permian to Cretaceous age, found scattered over volcanic in the area of Chitchun and Keogad in the north and northeastern part of Uttarakhand.

F 1 Fatehgarh Formation: A Paleocene to Eocene age fossiliferous formation of Barmer Basin, Rajasthan, consisting of conglomerate, sandstone, bentonitic clay, phosphatic sandstone and mudstone. See Table 16.

F 2 Fenestella Shale: This is a Carboniferous age formation of Lidder Valley, Kashmir sub-basin, overlying the Syringothyris Limestone and consisting of quartzite and shale rich in fossils of Fenestella, Productus, Spirifer, Pecten, etc. In the revised classification this formation is shown as Ganeshpur Formation. See Table 11.

F 3 Ferruginous Oolitic Formation: A Jurassic age formation of Malla Johar Supergroup in the Kumaon area of Uttarakhand. See the Supergroup and Table 17.

F 4 Frog Beds: These are Inter-trappean beds of Mumbai occurring in the Malabar, Cumballi, Worli, Antop, Sewri, Bhoiwada and Cotton Green Hills areas. The inter-trappeans consists of greenish ash beds, black shales and yellowish fine grained tuff material below basalt. Fossils of frog (Rana pusilla), fresh water tortoise (Carteremys leithii), cyprids, insects and fish tooth were obtained from the black shales. See Deccan Traps.

F 5 Fulra Limestone Formation: A Middle Eocene age formation of Kutch, Gujarat. See Table 38.

G 1 Gaj Formation: An Early Miocene age formation of sub-Himalaya region in Sindh-Salt Range area of Pakistan. See Table 22. In Saurashtra, Gujarat, Oligocene-Miocene age Ratanpur Clay, Bhumbli Conglomerate and Kuda Sandstone Members together are considered equivalent to Gaj Formation. See Table 20

G 2 Gajansar Beds: Jurassic age beds of Katrol Formation of Kutch, Gujarat. See Table 15.

G 3 Gamir Formation: A Precambrian Formation of Lesser Himalaya of Jammu and Kashmir. See Precambrian of Himalaya.

G 4 Ganeshpur Formation: See Fenestella Shale.

G 5 Gangamopteris Beds: These are plant fossil rich Permian age Lower Gondwana Beds of Jammu and Kashmir.

G 6 Gangapur Formation: A Gondwana Supergroup formation of Lr. Cretaceous age, occurring with Chikiala Formation in the Pranhita-Godavari Valley and consisting of sandstone, siltstone and clay. See Pranhita-Godavari Valley and Table 12.

G 7 Ganmachidam Formation: A Carboniferous age formation of Kanawar Group, overlying the Po formation in the Spiti-Kinnaur sub-basin of Himachal Pradesh and consisting of alternating pebbly mudstone, siltstone and sandstone and conglomeratic sandstone and black shale.

G 8 Garbyang Formation: The uppermost formation (probably of Silurian age) of Ordovician-Silurian succession, overlying the Shiala Formation in the Kumaon-Garhwal area of Uttarakhand.

G 9 Garhwal Group: A Precambrian Group of Lesser Himalaya in Uttarakhand. The Group includes Rautgara, Tejam, Berinag, Pithoragarh, Naraul and Bhowali Groups. See Precambrian of Himalaya.

G 10 Gechang Formation: A Lower Permian age Formation of Kuling Group of the Lingti Valley, Lalung village, Himachal Pradesh, consisting of conglomerate, sandstone, etc.

G 11 Geku Formation: See Yinkiang Group.

G 12 Ghuneri Member: See Umia Stage.

G 13 Giambal Gneiss: A Precambrian gneiss of Higher Himalaya in Himachal Pradesh. See Precambrian of Himalaya.

G 14 Girujan Clays/Formation: A Miocene-Pliocene age formation of Upper Assam Valley and Naga Hills Fold Belt in the Assam-Arakan Basin, consisting of soft mottled clays with sand and thin sandstone bands and lignite. See Table 23.

G 15 Giumal Formation: A Cretaceous age formation of Spiti area in Himachal Pradesh overlying the Spiti Formation and consisting of fossiliferous sandstone, siltstone, silty shale and shale with local conglomerate lenses.

G 16 Giumal Sandstone Formation: A Lower Cretaceous age formation of Malla Johar Supergroup, Kumaon area, Uttarakhand, consisting of glauconite sandstone, silt and radiolarians.

G 17 Godavari Supergroup: The sediments of Pranhita-Godavari Basin, Andhra Pradesh, which is one of the Proterozoic Sedimentary Basins forms the Godavari Supergroup, occurring in 2 parallel NW-SE trending sub-basins. The western sub-basin is called Pakhal or Mallampalli Belt and the eastern sub-basin is called Albaka Belt. See Table 30 in which formations of the two Belts are shown.

G 18 Godhara Granite: A Marwar Supergroup granite intrusion. It is a porphyritic granite to granodiorite with associated pegmatite and aplite and intrudes the Champaner and Lunavada Groups of the Aravalli Supergroup.

G 19 Gogunda Group: An associated group of South Delhi Fold Belt of the Delhi Supergroup which extends from Kishangarh near Ajmer in Rajasthan upto Himmatnagar in Gujarat and consists of muscovite-biotite schist, amphibolites, calc-silicate rocks and quartzite.

G 20 Golapalli Formation: A Lower Cretaceous age formation of Gondwana Supergroup in the Rajahmundry-Vijaywada area on the east coast in Andhra Pradesh, consisting of sandstone and grit with plant fossils. See Table 12.

G 21 Golden Oolite: A fossiliferous bed of Diadematus Zone of Macrocephalus Beds of Chari Formation in the Jurassic success in of Kutch, Gujarat. Golden Oolite consists of limestone whose rather coarse grains are coated with ferric oxide giving it golden colour. See Table 15.

G 22 Gondwana of Extra Peninsula: Formations having affinity with the Gondwana Supergroup are found at certain places in the Himalaya in the Extra Peninsular India. Important among these are Permian age Nishatbagh and Mamal Formations of Kashmir and Permian age Gondwana outcrops of Singrimari in Golpara district of Assam and in area between Kameng and Siang in Arunachal Pradesh.

G 23 Gondwana Supergroup: With its vast coal resources the Gondwana Supergroup becomes one of the most important supergroups in the Indian Geology. Formations of the Supergroup range in age from Upper Carboniferous/Lower Permian to Upper Jurassic/Lower Cretaceous and commonly consists of basal glacial boulder bed followed by sandstone, shale and coal. The Gondwana sediments were deposited in narrow intercratonic linear faulted basins e.g. Koel-Damodar, Son-Mahanadi and Pranhita-Godavari Basins. The Gondwana Supergroup can be best divided in to (a) Lower and (b) Upper Divisions, with line of division passing at the top of Panchet Formation. The Lower Gondwana Division is characterized by presence of Glossopteris plant fossils and Upper Division by presence of Ptilophyllum plant fosssils. On the east coast of India between Mahanadi and Cauvery upper Gondwana rocks of

Cretaceous and Tertiary age occur as isolated outcrops at Athgarh in Orissa, Ranasthalam, Eluru, Guntur, Ongole and Satyavedu in Andhra Pradesh and at Sriperumbadur and Uttatur in Tamil Nadu. These Upper Gondwana Coastal rock s are richly fossiliferous with pteridophytes, gymnosperms and invertebrates. See Table 12.

G 24 Goru Formation: A sub-surface formation of Cretaceous age of Jaisalmer Basin, Rajasthan encountered during drilling and consisting of shales with sandy lenses. See Table 16.

G 25 Gorubathan Formation: A Precambrian formation of Lesser Himalaya in Sikkim-Darjeeling area. See Precambrian of Himalaya.

G 26 Gorur Gneiss: The oldest Archean rock type (3300-3400 Ma) of the Western Dharwar Craton, consisting of trondhjemitic gneisses with associated tonalities and granodiorites. See Table 2.

G 27 Great Boundary Fault: A Fault zone of Aravalli Craton running NE-SW for about 400 km from near Chittorgarh to Sawai Madhopur in Rajasthan and separating Vindhyan Supergroup from the Hindoli Group- a formation of the pre-Aaravalli Basement sequence.

G 28 Grey Beds: Upper Triassic beds of Lilang Group, Spiti, Himachal Pradesh consisting of 150 m of fossiliferous grey shales and shaly limestone. See Table 13.

G 29 Grey Shale Formation: A Cretaceous formation of Trichinopoly area. See Trichinopoly area and Table 18.

G 30 Guitar Formation: A Pliocene age formation of Andaman-Nicobar Islands, consisting of calcareous mudstone. See Table 21.

G 31 Gund Formation: A Formation within Syringothyris Limestone Formation of Kashmir of Lower Carboniferous age.

G 32 Gundul Formation: See Abujhmar Group.

G 33 Gungri Formation: A Permian age formation of Kuling Group of Lahaul-Spiti-Kinnaur area in Himachal Pradesh, consisting of shale, phosphatic nodules and limestone.

G 34 Gurnaghat Formation: Lower formation of Khalsi Group of Late Cretaceous age in the Garo Hills area of Assam-Arakan Basin, NE India. See Table 23.

G 35 Gushal Formation: A formation of Tandi Group consisting of limestones and dolomite in southwest Lahaul in Himachal Pradesh of Late Paleozoic (Probably Permian) age. However fossils of gushal Formation indicate a Triassic age.

G 36 Gwalior Basin/Group: A Paleozoic Sedimentary Basin (Purana Basin) of Paleoproterozoic age in the Bundelkhand Craton, near Gwalior in Madhya Pradesh, NE of Bijawar Basin. The Gwalior Basin houses the Gwalior Group, which consists of 60 m thick lower Par Formation and 650 m thick upper Morar Formation. Par Formation consists of basal 'grit' and conglomerate grading into glauconitic sandstone. The Morar Formation consists of siliceous and ferruginous shale and chert, banded red jasper, limestone and tuffaceous sandstone. See Table 32.

H 1 Habur Formation: A Cretaceous age formation of Jaisalmer Basin in Rajasthan, consisting of coquinoidal limestone and calcareous sandstone. See Table 16.

H 2 Haimanta Group: A Precambrian to Lower Cambrian age Group of Higher Himalaya in Kashmir sub-basin, Uttarakhand and Himachal Pradesh. The Haimanta Group includes, Batal Formation at the base, followed by Kunzun La Formation (Lower and Upper). See Precambrian of Himalaya.

H 3 Halkal Formation: See Bhima Basin/Group.

H 4 Halobia Beds: Upper Triassic Beds of Lilang Group, Spiti, Himachal Pradesh, consisting of 45m of fossiliferous dark splintery limestone. Equivalent beds are also found in Byans and Painkhanda in Uttarakhand. See Table 13.

H 5 Halorites Beds: The term Halrites Beds is not in use now, the beds are part of Kioto Member of the Shal Shal Formation of Triassic (Norian) age in the Painkhanda area of Uttarakhand. See Table 26.

H 6 Hangrang Formation: An Upper Triassic formation of Lilang Group, Spiti, Himachal Pradesh, consisting of coral limestone bed. See Tables 13 and 14.

H 7 Hansera Group/Formation: See Bilara Group and Table 16.

H 8 Hapatnar Group: A Group of Cambrian to Mid. Silurian age in the Lidder Valley of Kashmir. See Table 11.

H 9 Harda Inlier: An Inlier of Bijawar Group, consisting of quartzite, breccias, limestone, dolomite, shale, sandstone and basic lava flows. See Bijawar Basin/Group and Table 32.

H 10 Harudi Formation: A middle Eocene age formation of Kutch, Gujarat. See Table 38.

H 11 Harwal Shale: See Bhima Basin/Group.

H 12 Hedenstoemia Beds: A Lower Triassic bed of Lilang Group, Spiti, Himachal Pradesh, consisting of 182 m of fossiliferous limestone, shaly limestone and shale. The formation extends to Painkhanda, etc. areas in Uttarakhand. See Table 13.

H 13 Higher Muschelkalk Member: A Member of Mikin Formation of Lilang Supergroup, Spiti, Himachal Pradesh. See the Supergroup and Table 14.

H 14 Hill Limestone Formation: A Paleocene-Eocene age formation of Potwar area, Jammu (J & K State). See Table 22.

H 15 Himgiri Beds: Lower Gondwana beds of Raigarh-Himgiri coal fields of Chhattisgarh, equivalent to Raniganj Formation.

H 16 Himmatnagar/Ahmadnagar Sandstone: A massive and horizontally bedded sandstone formation with shale of Lower Cretaceous age in the Himmatnagar area of Gujarat.

H 17 Hindoli Group: The Archean age oldest Group of Bhilwara Supergroup of Aravalli Craton, named after village Hindoli, near Bundi in Rajasthan, consistis mainly of greywackes and phyllites with some cross-bedded quartzite, dolomitic limestone and algal dolomite. See Table 6.

H 18 Hirpur Formation: A Pliocene age formation of Karewa Group of Jammu and Kashmir. See Table 24.

H 19 Hosakatti Argillite: See Kaladgi Basin and Table 37.

H 20 Hoti Member: A Member of Shal Shal Formation of Triassic age in Painkh-anda area of Uttarakhand, consisting of limestone and shale. See Table 26.

H 21 Hundri Formation: The formation represents the Cretaceous rocks of Eastern Korakoram area of Jammu and Kashmir State and lies in the Shyok Tectonic Belt/Syyok Suture and consists of massive limestone in the lower part and shales in the upper part.

H 22 Hydrocarbons producing basins of India: See Offshore Basins and Table 29.

i 1 Ichnology: It is the study of trace-fossils or ichno fossils, especially the study of fossil tacks. Trace fossils are traces of horizontal burrows, trails and resting excavations. Labenspurren are the sedimentary structures left by a living organism. Stromatolites are best examples of trace fossils. Stromatolites are not 'body fossils' but traces of life activity. These are sedimentary structures built by bacteria, cyanophytus and microscopic algae, and only rarely do they contain remains of organisms that formed them.

i 2 INDO-GANGATIC PLAIN: it is a depression between the Himalaya and the Indian Peninsula, extending from Punjab to West Bengal. It is filled with river borne unconsolidated quaternary sedimentary deposits of Older Alluvium and Newer Alluvium consisting mainly of sand, silt and clay to form a present day nearly flat and plain stretch of land. The Older Alluvium is of Middle Pleistocene age and predominantly consists of well oxidised clay, silt and sand with calcareous/ferruginous nodules/concretions/kankar, etc., while Newer Alluvium is mostly of Upper Pleistocene to Holocene (Recent) age and predominantly consist of unoxidised and pale light coloured clay, silt and sand. The Indo-Gangatic Plain on north is separated from the Himalaya by the Main Boundary Fault.

i 3 INDRAVATI BASIN: age of the Bastar Craton, lying in the Bastar area around Jagadalpur in the south-eastern part of Madhya Pradesh. See Tables 32 & 36.

i 4 INDUS FORMATION: A paleocene age formation of upper Sindhu valley in the sub-himalayan region. see Table 22.

i 5 INFRA-TRAPPEAN BEDS : Thes are sedimentary fossiliferous horizons lying below the deccan traps. See deccan traps. Fossiliferous infra-trappean (=lameta) beds occur at the following localities: 1.Andhra Pradesh -a. Marepalli, b. Rajahmundry; 2. Gujarat- a. Rahioli; 3. Madhya Pradesh- a. Jabalpur; 4. Maharashtra - a. Dongargaon, b. Pisdura, c. Umrer; 5. Subsurface infra-trappean beds encountered in the Krishna-Godavari Basin (kg) at- a.

Chintalapalli-a, b. Elamanchili-a, c. Modi-a, d. Narsarpur well-1, and e. Palakollu.

i 6 INGLIS FORMATION: A formation of Miocene age of Andaman and Nicobar Islanbleds. see Table 21.

i 7 INOLGIRI MEMBER: Upper member of Kherapara formation, equivalent to Barail group of oligocene age, occurring in the garo hills area of Assam-Arakan basin in NE India. see Table 23.

i 8 INTER-TRAPPEAN BEDS: These are fossiliferous sedimentary horizons found in between the different layers of deccan traps. see deccan traps. fossiliferous inter-trappean beds are found at the following localities: 1. Andhra Pradesh - a. Asifadad, b. Rajahmundry, c. Naskal, d. Rangapur; 2. gujarat- a. Anjar, b. Bamanbor, c. Kora, d. Lakshmipur; 3. Karnataka- a. Biligi, b. Gurmatkal, c. Upparhatti, d. Yellur; 4. Madhya Pradesh- a. Kisalpuri, b. Mohgaon-Kalan, c. Padwar/Ranipur, d. Mandla, e. Sagar; 5. Maharashtra- a. Bombay (Mumbai), b. Nagpur; 6. Rajasthan - a. Mamoni; 7. Uttar Pradesh- a. Lalitpur; 8. Krishna-Godavari Basin (fossils encountered subsurface) at- a. Chintalapalli-a, b. Elamanchili-a, c. Modi-a, d. Narsarpur well-1, e. Palakollu-a, f. Razole-a; and 9. Rajmahal basin in new part of Jharkhand. intertrappean beds are also found at 10. Chimalgutta and 11. Mallangutta.

J 1 Jabalpur Formation: A gondwana supergroup formation of lower cretaceous age in the satpura basin. the formation consists of conglomerate, clay and sandstone. see Table 12.

J 2 Jagannath or Dungoaposi Volcanics: A volcanic group of the Singhbhum Craton younger than Singhbhum Granite and Koira group and older than Kolhan Group and consisting of 30 volcanic flows which consists of komatiitic and tholeiitic metabasalts with minor basaltic andesites and basanites. See Table 5.

J 3 Jagat Group: Fragments of ancient rocks like amphibolites, quartzite, mica schist, crystalline limestone (marble), calc-silicate rocks and banded iron formation, found scattered within the Mewar Gneiss, in Jagat area, near Udaipur form the Jagat Group. See Table 6.

J 4 Jagdalpur Formation: See Indravati Basin

J 5 Jaintia Group: The Group represents Paleocene to Eocene rocks of Assam-Arakan Basin of NE India. Disang Formation, Rewak Formation, Kopili Formation and Sylhet sub-Group are equivalent to Jaintia Group. See Tables 22 and 23.

J 6 Jaisalmer Formation: A Middle Jurassic age formation of Jaisalmer and Barmer Basins of western Rajasthan, consisting of fossiliferous calcareous sandstone and limestone with Oolitic bed at the top. See Table 16.

J 7 Jakaram Arkose Formation: Oldest formation of Godavari Supergroup. See Godavari Supergroup and Table 30.

J 8 Jalore Granite: A Malani Igneous Suite Granite consisting of 7 plotons, among which Sankara pluton is the largest, covering 3700 sq.km. area and consisting of rhyolite, granite and felsic to mafic dykes. The Granite is mostly pink in colour with bands and patches of grey granite in it.

J 9 Jambusar Formation: A Pleistocene age sub-surface formation of Mainland, Gujarat. See Table 20.

J 10 Jaunsar Group: A Precambrian Group of Lesser Himalaya in Himachal Pradesh. The Group includes Peontra Volcanics. See Precambrian of Himalaya.

J 11 Jenam Formation: An Oligocene age formation, equivalent to the Middle Part of Barail Group and consisting of shales and sandstones and occurring in the Khasi and Jaintia Hills, Mikir Hills and Dhansiri Valley and Surma Valley and Tripura Fold Belt areas of Assam-Arakan Basin in NE India. See Tables 22 and 23.

J 12 Jhagadia Formation: A Middle Miocene to Lower Plicene age formation of Gujarat Mainland, consisting of 200 m. of unaossiliferous light grey sandstone and pebble to cobble conglomerate. The formation overlies Kand Formation. See Table 20.

J 13 Jhamarkotra Formation: See Debari Group.

J 14 Jhangu Formation: A Formation of Upper Cretaceous age of Malla Johar Supergroup, Uttarakhand. See the Supergroup and Table 17.

J 15 Jharol Group: A Group of Aravalli Supergroup, which fans out to the south from Nathdwara for a distance of 200 km. with 40 km. width and consists of carbonate free pelites and arenites with conglomerate bands at the base. See Table 6.

J 16 Jodhpur Group/Formation: Lowermost Group of Nagaur Basin of Marwar Supergroup of Neoproterozoic age in the Aravalli Craton, consisting of sandstone, shale, limestone and chert. Now the Marwar Supergroup has been assigned Permo-Carboniferous age instead of Proterozoic-Cambrian assigned earlier. See Marwar Supergroup.

J 17 Jogira Formation: A fossiliferous Eocene age formation of Bikaner-Nagaur Basin in western Rajasthan, overlying the Marh Formation and consisting of limestone, marl and fullers earth. See Table 16.

J 18 Jurassic of Kashmir: See Drumgam Formation.

J 19 Jurassic of Kutch: Comprises of Pachcham, Chari and Katrol Formations. See each formation.

J 20 Jurassic of Painkhanda: Consists of Lapthal Member of Kiogad Formation. See Table 26.

J 21 Jutogh Group: A Precambrian Group of Higher Himalaya in Himachal Pradesh. See Precambrian of Himalaya.

J 22 Juvavites Beds: An Upper Triassic bed of Lilang Group of Spiti, Himachal Pradesh, consisting of 150 m of fossiliferous shales, limestones and sandstones. See Table 13.

K 1 Kaga Formation: A Middle Triassic formation of Lilang Group, Spiti, Himachal Pradesh, consisting of Daonella Shales. See Tables 13 and 14.

K 2 Kailas (Kargil) conglomerate: An Eocene to Early Miocene age formation of Upper Sindhu Valley in the sub-Himalayan region of Jammu and Kashmir. See Table 22.

K 3 Kaimur Group: 400m thick second lowest Group of the Vindhyan Supergroup. See Table 9.

K 4 Kakara Formation: Paleocene to Lower Eocene age formation of Lesser Himalayan tectogene in Himachal Pradesh, consisting of basaltic pisolitic laterite/bauxite followed by quartz arenite, variegated shales and massive to thin bedded limestone.

K 5 Kakdi Stage: Lower stage of Early Eocene age, of Berwali Series of Kutch, Gujarat, also encountered in Lakhpat I well of Kutch, Gujarat. See Tables 20.

K 6 Kaladgi Basin: A Purana Proterozoic (Meso-Neoproterozoic) Sedimentary Basin of Dharwar Craton. The Basin is occupied by lower Bagalkot Group, consisting of Lokapur and Simi keri sub-Groups and upper Badami Group, consisting of lower Kerur Arenite and upper Katageri Limestone. See Tables 32 & 37.

K 7 Kalamund Formation: A Precambrian formation of Lesser Himalaya in Jammu and Kashmir. See Precambrian of Himalaya.

K 8 Kalapani Limestone: A Triassic age formation of northern Kumaon in Byans, Uttarakhand, consisting of fossiliferous limestone succeeding the Chocolate Series.

K 9 Kalhal Formation: A Triassic age formation overlying the Permian age Salooni formation of Chamba area in Himachal Pradesh, consisting of limestone and dolomite with interbedded quartzite in the upper part.

K 10 Kalkel Formation: A Trassic age formation overlying the Permian age Salooni Formation in the Chamba sub-basin of Himachal Pradesh.

K 11 Kallakkudi Limestone Formation: See Trichinopoly area and Table 18.

K 12 Kallamedu Formation: See Trichinopoly area and Table 18.

K 13 Kallankurichi Formation: See Trichinopoly area and Table 18.

K 14 Kamakhyanagar Quartzite: A formation of Singhbhum Craton, correlated with Kohlan Group and consisting of cross bedded and cherty quartzites bounded by faults on either side. See Table 5.

K 15 Kamlial Formation: See Siwalik Group and Table 28.

K 16 Kamthi Formation: A Gondwana Supergroup formation of Upper Permian to Lower Triassic age, equivalent to Raniganj and Panchet Formations, occurring in the Son-Mahanadi and Godavari Valley areas. In the Son-Mahanadi area Kamthi Formation consists of sandstones and shales, while in the Godavari Valley area it consists of conglomerate, conglomeratic sandstone, siltstone and shale. See Pranhita-Godavari Valley and Table 12.

K 17 Kanawar Group: Represents the Carboniferous formations of Spiti-Kinnaur sub-basin of Himachal Pradesh and Kumaon sub-basin of Uttarakhand. In Spiti-Kinnaur sub-basin the Kanawar Group consists of Lipak, Po, and Ganmachidam Formations, while in Kumaon sub-basin the Kanwar Group consists of Lipak and Po Formations.

K 18 Kanchanjhanga Gneiss: A Precambrian gneiss of Higher Himalaya in Sikkim-Darjeeling area. See Precambrian of Himalaya.

K 19 Kand/Kond Formation: A Lower Miocene age formation of Gujarat Mainland, consisting mainly of calcareous sandstone with occasional conglomerate, clay, marl and thin fossiliferous limestone. See Table 20.

K 20 Kanger Formation: See Indravati Basin.

K 21 Kankawati Series: A Pliocene age series of Kutch, Gujarat, overlying the Khari Series and consisting of fossiliferous micaceous quartzose sandstone in the lower part, conglomerate in the middle part and grits and sandstones with calcareous nodules in the upper part. See Tables 20 and 38.

K 22 Kanker Granitoids: A Proterozoic age granite that intrudes Sonakhan Belt of Bastar Craton. See Table 4.

K 23 Kanthkot Sandstone: A Jurassic age formation of Kutch, Gujarat. See Table 15.

K 24 Kaptipada granite: A Singhbhum Craton Archean granite having xenoliths of migmatites and amphibolites and is rich in Al_2O_3 trondhjemite-granodiorite. See Table 5.

K 25 Kapurdi Formation: A fossiliferous Eocene age formation of Barmer Basin in western Rajasthan, consisting of fullers earth with ferruginous and gypsiferous concretions and bioclastic limestone. See Table 16.

K 26 Karai Shale Formaion: A Cretaceous age formation of Uttatur Group of Trichinopoly area, Tamil Nadu. See Table 18.

K 27 Karal: These are beach rocks of middle to Late Holocene age occurring as isolated patches along coast in Maharashtra, consisting of conglomerate, shell-limestone bands and consolidated dune sand. The beach rocks contain fossils of mollusca, foraminifera, broken fragments of bryozoa, corals, etc. See Quaternary sediments of Maharashtra.

K 28 Karewa Group: The Group includes Pliocene to Pleistocene formations of Karewa Basin in Kashmir Valley. For details see Table 24.

K 29 Kargil Conglomerate: See Kailas Conglomerate.

K 30 Karharbari Formation: Gondwana Supergroup's second oldest formation of Permian age, overlying the Talchir Formation and consisting of conglomerate, sandstone, shale and thin coal bands. See Table 12.

K 31 Karihul Formation: A Precambrian formation of Higher Himalaya of Jammu and Kashmir State. See Precambrian of Himalaya.

K 32 Karikal Beds: In 1884 in an artisian well boring at Karikal (Karai Kal) near east coast in Tamil Nadu, between 58 and 105 m depth 7-8 horizons consisting of clay, sand, pebbles and conglomerate were found rich in Tertiary (most probably Pliocene) fossils which included about 100 species of Gastropods as identified by Crossmann.

K 33 Karruchal Formation: A Late Mesozoic formation of Kerala, consisting of hard gritty sandstone and conglomerate near base and clay interbeds with arkose in the upper part.

K 34 Kasauli Formation: The uppermost formation of Upper Oligocene to Lower Miocene age of Sirmur Group in the Lesser Himalaya of Himachal Pradesh consisting of 1250 m of massive sandstone and subordinate shale. See Table 22.

K 35 Katageri Limestone: See Kaladgi Basin and Table 37.

K 36 Katamdevarahalli Limestone: See Bhima Basin/Group.

K 37 Katesar Member: See Umia Stage.

K 38 Katrol Formation: A Jurassic Formation of Kutch, Gujarat. For details see Table 15.

K 39 Kekri Formation: See Mangalwar Comples and Table 6.

K 40 Kerur Arenite: See Kaladgi Basin and Table 37.

K 41 Khairagarh Group: A Proterozoic group of Kotri-Dongarhgarh Orogen in Bastar Craton and consists of from bottom to top of Bortalao Sandstone, Sitagota Basalt, Karutola Sandstone, Mangikhuta Basalt and Ghogra Sandstone. See Table 4.

K 42 Khalsi Formation: See Sangeluma Group.

K 43 Khari Nadi Formation: Lower formation of Miocene age Khari Series in Kutch, Gujarat. See Table 38.

K 44 Khari Series; A Miocene age series of Kutch, Gujarat. See Tables 20 and 38.

K 45 Khariar Basin: It is one of the Proterozoic Sedimentary Basins (Purana Basin) of Meso-Neoproterozoic age in the Bastar Craton. The sediments of Khariar Basin form the Pairi Group, which is 600 to 1000 m thick. The Basin lies in the eastern part of Madhya Pradesh, adjoining the Orissa Border and west of Bhawanipatna and Sambalpur. See Tables 32 and 35.

K 46 Khasil Group: Late Cretaceous age Group of Assam-Arakan Basin. In Garo Hills the Group consists of lower Gurmaghat and upper Mahdek Formations, In Khasi and Jaintia Hills the Group consists from bottom to top of Borghat, Mahadek and Langpar Formations, while in Mikir Hills and Dhansiri Valley the Group is represented by Dergaon Formation. See Table 23.

K 47 Kherapara Formation: A formation of Oligocene age, equivalent to Barail Group, occurring in the Garo Hills area of Assam-Arakan Basin in NE India. The formation consists of lower the Darik Member and upper the Inolgiri Member. See Table 23.

K 48 Khetri Basin: Occurs in the North Delhi Fold Belt of the Delhi Supergroup. The basin is famous for its copper deposits and is divided into North Khetri Basin (NKB) and South Khetri Basin (SKB), separated by Kantli Fault. NKB includes Khetri Group which consists of ferruginous breccias, quartzite, amphibolites-biotite quartzite, para amphibolites, impure marble, garnet-chlorite schist, andulasite-biotite schist and metagreywacke. Several granite bodies e.g. Gotri, Seoli and Jasrapur granites are present within the NKB. SKB includes Shyamgarh Group which consists from bottom to top of staurolite-garnet schist with pebbly quartzite, felsic volcanic and pyroclastics, carbon phyllite and chert, conglomeratic quartzite, garnet-mica schist, Banded Iron Formation (BIF), dolomite, calc-silicate rock, semipelitic schist and greywacke, graphitic schist and micaceous quartzite. Like NKB, the SKB is also intruded with many granitic bodies, e.g. Paniaras granite, Chhapoli granite, Udaipurwati granite and Sladipura granite. See Table 8.

K 49 Khreuh Formation: A Middle Triassic age formation of central Kashmir valley, Pastannat area, consisting predominantly of sandy limestone and shale with little limestone.

K 50 Khuiala Formation: An Eocene age fossiliferous formation of Jaisalmer Basin in western Rajasthan, consisting of limestone and shale. See Table 16.

K 51 Khunamuh Formation: A Permian to Lower Triassic age formation of Kashmir, Laddakh and Eastern Korakoram in J & K, overlying the Permian age Zewan Formation and consisting of limestones with fossils of Claria (a lamellibranch), etc.

K 52 Kimberlites: Diamond bearing and non-diamond bearing kimberlite diatremes pipes have been found associated with the Proterozoic Sedimentary Purana Basins of Peninsular India. Important among these pipes are Majhgawan and Hinota pipes in the Vindhyan Basin.

K 53 Kiogad Formation: A Jurassic age formation of Niti area, Painkhanda, Uttarakhand, consisting of fossiliferous shales with calcareous nodules. The formation comprises of 3 Members-Spiti, Chhojan and Lapthal Members.

K 54 Kioto Formation/Group: The uppermost formation of Lilang Group in the Triassic succession of Spiti, Himachal Pradesh and Kumaon, Byans, Uttarakhand, consisting of Megalodon Limestone (Para Stage) forming the lower bed and Tagling Limestone forming the upper bed. When Lilang is considered as a Supergroup, Kioto becomes a Group. See Tables 13 and 14.

K 55 Kioto Member: Uppermost Member of Triassic age of Shal Shal Formation of painkhanda area, Uttarakhand, consisting of limestone and quartzites. See Table 26.

K 56 Kirthar Formation: Eocene age formation of Sindh-Salt Range area of Pakistan. See Table 22.

K 57 Kohlan Group: One of the younger groups of Singhbhum Craton, occurring in Chaibasa – Noamundi area. The group consists of 3 formations which from bottom to top are Mungra Sandstone, Jhinkpari Limestone and Jetia shale. The Mungra Sandstone is about 25m thick and consists of conglomerate and sandstone. The Jhinkpari Limestone is about 20m thick and consists of limestone followed by shale. The Jetia Shale is thickest of all, consisting of about 200m of purple shale, calcareous shale and limestone. See Tables 5.

K 58 Koira Group: A group of Singhbhum Craton comprising of 5 formations with a basal polymict conglomerate overlain by a sequence of quartz arenite and shale with local dolomite and Banded Iron Formation and is intruded by gabbro and dolerite. Presence of sedimentary structures like ripple marks, current bedding etc. indicate near shore depositional conditions. See Table 5.

K 59 Kolar Group: Upper Group of Dharwar Supergroup in the Eastern Dharwar Craton, overlying the Yashwantanagar Formation. The Group consists form bottom to top of Sakarsanthall Formation, Gold Field Volcanics and Champion Gneiss. See Tables 2 and 3.

K 60 Kondagaon Granulites/Belt: Is about 70 km long and is of Neoarchean age and exists in the midst of Bastar Craton. Charnockites and leptynites dominate in the western part of the Belt, while in the eastern part pillow lavas of Bengpal Group are present. See Table 4.s

K 61 Kopili Formation: Eocene age, uppermost formation of Jaintia Group in the Khasi and Jaintia Hills, Mikir Hills and Dhansiri Valley and Upper Assam Valley of Assam-Arakan Basin of NE India. See Tables 22 and 23.

K 62 Kota Formation: A Jurassic age formation of Gondwana Supergroup in the Godavari Valley area, consisting of sandstone with Glossopteris Flora in the lower part, limestone with fish and dinosaur fossils in the middle part and sandstone and clay with Ptilophyllum flora in the upper part. See Pranhita-Godavari Valley and Table 12.

K 63 Krishna-Godavari Basin: See Offshore Basins and Table 29.

K 64 Krol Belt/Formation/Group: A Precambrian formation of Lesser Himalaya in the Nainital syncline of Uttarakhand. See Precambrian of Himalaya.

K 65 Kuar Bet Beds: Same as Pachcham basal beds.

K 66 Kuda Member: Miocene age uppermost member of Gaj Formation in Saurashtra, Gujarat, consisting of sandstone. See Table 20.

K 67 Kuilapal Granite: A gneissic granite of Younger Granitoids of the Singhbhum Craton ranging in composition from trondhjemite to granite. See Table 5.

K 68 Kukti Formation; The lowermost formation of Tandi Group of Permian age in Himachal Pradesh, consisting of phyllites, limestone, calcarenite and pebble beds with Permian age fossils.

K 69 Kulakkalnattam Formation: See Trichinopoly area and Table 18

K 70 Kuling Shale/Formation/Group: A Permian age formation of Niti area, Painkhanda, Uttarakhand, consisting of fossiliferous sandstone, limestone and shale with nodules. Kuling Formation is also exposed in Mulbekh and Lamayuru areas of Jammu and Kashmir in close association with Lilang Group of Triassic age.

K 71 Kullu Group: A Precambrian Group of Higher Himalaya in Himachal Pradesh. See Precambrian of Himalaya.

K 72 Kundargi Quartzite: See Kaladgi Basin and Table 37.

K 73 Kunzun La Formation: Upper formation of Haimanta Group of Precambrian to Lower Cambrian age in the (a) Spiti area of Himachal Pradesh and (b)

Zanskar area of Jammu & Kashmir. See Haimanta Group and Precambrian of Himalaya.

K 74 Kurat Basalt: See Sonrai Basin.

K 75 Kurnool Group: A Proterozoic age Group of Cuddapah Basin overlying the Cuddapah Supergroup, consisting of 2 Subgroups-lower the Jammalamaduga Subgroup and upper the Kundair Subgroup. The Jammalamaduga Subgroup comprises of 3 formations, which from bottom to top are - 1. Banganpalle Quartzite consisting of oligomict conglomerate, grit, quartzite and shale; 2. Narji Limestone consisting of flaggy grey limestone and glauconitic sandstone; and 3. Auk Shale consisting of laminated shale with siltstone interbeds. The Kundair Subgroup also comprises of 3 formations which from bottom to top are - 1. Paniam Quartzite, consisting of pinnacled quartzite, siliceous shale and basal pebble bed; 2. Koikuntla Limestone, consisting of siliceous shaly limestone with quartzite interbeds and 3. Nandyal Shale, consisting of shale partly calcareous. See Table 2.

K 76 Kuti Shales: A Triassic age formation of northern Kumaon in Byans, Uttarakhand, consisting of fossiliferous micaceous shale with calcareous flagstone.

L 1 Labenspurren: See Ichnology.

L 2 Laisong Formation: The lowermost formation of Oligocene age, equivalent to the lower part of the Barail Group, consisting of sandstones and shales and occurring in Khasi and Jaintia Hills, Mikir Hills and Dhansiri Valley and Surma Valley and Tripura Fold Belt areas of Assam-Arakan Basin in the NE India. See Tables 22 and 23.

L 3 Lakadong Limestone Member: Lowermost Member of Sylhet Subgroup of Jaintia Group of Eocene age in Khasi and Jaintia Hills of Assam-Arakan Basin in NE India. See Tables 22 and 23.

L 4 Lakadong Sandstone Member: A Sylhet Subgroup Member of Jaintia Group of Eocene age, overlying the Lakadong Limestone Member in the Khasi and Jaintia Hills of Assam- Arakan Basin of NE India. See Table 23.

L 5 Lakanavaram Shale Formation: A formation of Godavari Supergroup. See Godavari Supergroup and Table 30.

L 6 Laki Formation: See Table 22.

L 7 Lalab Formation: A Lower Cambrian age formation of Kashmir Tethyan sub-basin, consisting predominantly of laminated slates and quartzose partings or bands.

L 8 Lamayuru Formation: A Hettangian (lowermost Jurassic) age formation of Ladakh, J & K State, consisting of a clastic facies (shale/siltstone).

L 9 Lameta Formation: An uppermost Cretaceous (Maastrichtian) age formation just younger to Bagh Formation and occurs as detached outcrops in parts of central and western India in Balasinor-Jhabua, Jabalpur, Sagar, Ambilakur-Amarkantak and Nand-Dongargaon Basins with type area Lameta, 15 km SE of Jabalpur. Lameta are both Infra-trappean and inter-trappean beds of the

Deccan Tasps. Lameta Formation predominantly consists of varities of limestones and sandstones and contain fossils of aquatic, semi-aquatic and terrestrial fauna and flora, which include Dinosaur egg shells also.

L 10 Lamprophyre: Are included among the granite intrusions of Marwar Craton of the Aravalli Craton and occur as dykes and correlates with the alkaline magmatism of the Deccan Traps.

L 11 Langarin Member: It underlies Sylhet Subgroup of Jaintia Group in the Garo Hills of Assam-Arakan Basin of NE India. See Table 23.

L 12 Langpar Formation: Uppermost formation of Khalsi Group of Late Cretaceous to Paleocene age in Cherra Plateau Meghalaya and Khasi and Jaintia Hills of Assam-Arakan Basin of NE India. See Table 23.

L 13 Lapthal Member: Lowermost member of Kiogad Formation of Jurassic age in Niti area of Kumaon, Painkhanda, Uttarakhand, consisting of thinly bedded limestone. See Table 26.

L 14 Largi Group: A Precambrian Group of Lesser Himalaya in Himachal Pradesh. See Precambrian of Himalaya.

L 15 Largi Window: A structure in the second Precambrian sedimentary cycle of Lesser Himalaya in Himachal Pradesh, represented by older Rampur Group and younger Largi Group.

L 16 Lassaria Formation: See Mangalwar Complex and Table 6.

L 17 Laterite: Laterite is derived by intense subaerial weathering of igneous and sedimentary rocks in Peninsular India, forming low mounds and are rich in iron and aluminum and have rusty red colour and are used as building stone. For laterite there is a National Geological Monument at Angadipuram in Kerala.

L 18 Lathi Formation: A Jurassic (Lias) age formation of Jaisalmer and Barmer Basins in western Rajasthan, comprising of Thaiat and Odania Members and consisting of fossiliferous sandstone, ferruginous siltstone, purplish white clay and conglomerate. See Table 16.

L 19 Lidder Group: A Permian-Carboniferous Group of Lidder Valley, Kashmir.

L 20 Lignite Deposits: See Neyveli Lignite Field and Palana Formation.

L 21 Lilang Group: Fossiliferous Triassic-Jurassic succession of Spiti, Himachal Pradesh is represented by Lilang Group. The Group overlies Permian Products Shale and underlies Jurassic Spiti Shale Formation. See Table 13.

L 22 Lilang Supergroup: In an alternative arrangement the Triassic–Jurassic rocks of Spiti, Himachal Pradesh have been put under Lilang Supergroup. See Table 14.

L 23 Limestone and Shale Member: A Member of Mikin Formation of Tamba Kurkur Group of Lilang Supergroup, Spiti, Himachal Pradesh. See Table 14.

L 24 Lipak Formation: Lowermost Carboniferous age formation of Kanawar Group of Spiti-Kumaon and Kumaon sub-basin. The formation consists of sandstone in the lower part, shale interbedded with limestone and sandstone in the Middle part and limestone, shale and gypsum bed in the upper part.

L 25 Lochambal Beds: Uppermost beds of Spiti Shale Formation consisting of shale, limestone and sandstone with fossils of ammonite, lamellibranchs and belemnites. In the current literature the name Lochambal Beds may not find any mention and are treated as part of the Spiti Shale Formation.

L 26 Lokapur Subgroup: See Kaladgi Basin.

L 27 Lolab Formation: A Precambrian to Lower Cambrian age formation of Higher Himalaya in Jammu and Kashmir, consisting of predominantly of laminated slates and quartzose partings or bands. See Precambrian of Himalaya.

L 28 Long Formation: A Miocene age formation of Andaman and Nicobar Islands. See Table 21.

L 29 Lower Katrol: A Jurassic age bed of Katrol Formation of Kutch, Gujarat. See Table 15.

L 30 Lower Limestone Member: Lower Triassic age Member of Mikin Formation of Lilang Supergroup, Spiti, Himachal Pradesh. See Lilang Supergroup and Table 14.

L 31 Lower Muschelkalk. A Triassic age bed of Mikin Formation of Lilang Group, Spiti, Himachal Pradesh. See Lilang Group and Table 13.

L 32 Lumla Formation: A Precambrian formation of Higher Himalaya in Arunachal Pradesh. See Precambrian of Himalaya.

L 33 Lunavada Group: A Group of Aravalli Supergroup, consisting predominantly of greywacke-phyllite with associated quartz arenites. Minor bands of dolomite, phosohatic and stromatolitic solomite, manganiferous phyllite and polymict conglomerate are also present. See Aravalli Supergroup and Table 6.

M 1 Macchal Formation: A Precambrian formation of Lesser Himalaya in Jammu and Kashmir. See Precambrian of Himalaya.

M 2 Macrocephalus Beds: Lowermost beds of Chari Formation in the Jurassic succession of Kutch. See Table 15.

M 3 Madh Series: A Paleocene age stage of Kutch, Gujarat, consisting of tuffs, volcanic ash, tuffaceous shale, sandstone, bentonitic clay and lignite deposits at Panandro. The formation overlies the Deccan Traps and contain fossils of Dicot leaves impressions, spores and pollens indicating Paleocene-Eocene age for the stage. See Table 20.

M 4 Mahadek Formation: A formation of Khạlsi Group of Late Cretaceous age occurring in Garo Hills, Khasi and Jaintia Hills of Assam-Arakan Basin of NE India. See Table 23.

M 5 Mahadeo Formation: A 50 m thick formation of Late Cretaceous to Early Paleocene age of Meghalaya Group in Meghalaya, NE India. See Meghalaya Group.

M 6 Mahadeva Formation: The term is sometimes used to collectively include Pachmarhi, Denwa and Bagra Formations of Gondwana Supergroup.

M 7 Mahakoshal Supracrustal Belt/Group: A Proterozoic formation of Satpura Mobile Belt bound on north by Son Narmada North Fault and on south by Son Narmada South Fault and consisting of quartzite, carbonate, chert, Banded

Iron Formation, greywacke-argillite and mafic volcanic. The Mahakoshal Supracrustal Belt/Group from bottom to top consists of Sleemanabad Formation, Parsoi Formation and Dudhamaniya Formation. In another classification Mahakoshal Group has been divided from bottom to top into Agori Formation and Chitrangi Formation (equivalent to Sleemanabad Formation) and Parsoi Formation, which includes both the Parsoi and Dudhamaniya Formations.

M 8 Mahakut Breccia: See Kaladgi Basin and Table 37.

M 9 Main Boundary Fault: Separates the Indo-Gangatic Plain from the Himalaya which are mostly Siwaliks .

M 10 Main Central Thrust: Separates the Lesser Himalaya from the Higher Himalaya.

M 11 Malabar Supergroup: The Tertiary and Quaternary Formations of Kerala are together grouped to form the Malabar Supergroup, which includes Tertiary Warkalli Group and Quaternary Vembanad Formation. See the Warkalli Group and Vembanad Formation and Table 25.

M 12 Malangtoli (Nuakot) Volcanics : An younger volcanic group of the Singhbhum Craton, lying in association with Kolhan and other equivalent Groups and ranges in composition from basalts to basaltic andesites. See Table 5.

M 13 Malani Granite: Belongs to Malani Igneous Suite and includes two plutons of hornblende granite, consisting of porphyritic granite rich in hornblende and grades into alkali granite porphery, covering 70 sq. km near Barmer in western Rajasthan.

M 14 Malani Igneous Suite: Occurs in the Marwar Craton of Western Rajasthan and forms a large Volcanic Province, covering about 51,000 sq.km area. The Suite occurs as scattered outcrops in the form of ring intrusions and radial dykes in the otherwise plain and undulated area of Thar Desert covered with wind blown sand and sand dunes. The Suite consists of rhyolite, perlite and trachyte associated with volcaniclastics like tuffs and breccias. The Suite includes Siwana granite, Jalore granite, Tusham granite, Jhunjhunu granite and also the Kirana granite of Pakistan. The Suite overlies the Sirohi and Sindreth Groups and younger granitoids of Delhi Supergroup, particularly the Erinpura and Sendra-Ambaji granite suites and pre-Delhi gneisses. See Tables 6 and 16.

M 15 Malanjkhand Granite: One of the younger granites that intrudes the Bastar Craton. Malanjkhand copper deposit is one of the largest copper deposits in India. See Table 4.

M 16 Maleri Formation: A Gondwana Supergroup formation of Triassic age of Godavari valley, consisting of clay, siltstone, argillaceous sandstone and lime-pellet rock. The formation from bottom to top is divisible into Yerapalli, Bhimaram, Maleri and Dharmaram Members. These Members are also considered as Formations. See Pranhita-Godavari Valley and Table 12.

M 17 Malla Johar Supergroup: The Jurassic-Cretaceous rocks of Kumaon Himalaya of Malla Johar area near Soncha Malla in Pithoragarh district of Uttarakhand have been grouped under Malla Johar Supergroup. See Table 17.

M 18 Mallampalli Group: Oldest Group of Godavari Supergroup of Proterozoic age in Pakhal or Mallampalli Belt. See Godavari Supergroup and Table 30.

M 19 Mamal Formation: A Lower Permian age formation containing plant fossils and consisting of novaculite, limestone, tuffaceous shale, carbonaceous shale, purple and pinkish shale with arenite, overlying the Punjal Volvanics in the Kashmir Sub-basin.

M 20 Mancharial Quartzite Formation: A formation of Sullavai Group of Pakhal or Mallampalli Belt of Godavari Supergroup. See Godavari Supergroup and Table 30.

M 21 Mandi Volcanics: Volcanics associated with the Sundergarh Group of Precambrian sedimentary cycle II of Lesser Himalaya in Himachal Pradesh.

M 22 Manendragarh Marine Bed: Like the Umaria Marine Bed, a 5 m thick marine bed of Permian age occurs under the railway bridge on Hasdeo river, about 4 km SW of Manendragarh railway station, and about 150 km ESE of Umaria in Madhya Pradesh. The Manendragarh Marine Bed occurs associated with the lower part of the Talchir Formation of the Gondwana Supergroup, and consists of various types of shales, conglomerate and green sandstone at the top. Fossils are found in the conglomerate horizon at the base of the marine bed.

M 23 Mangalwar Complex: It is the middle formation of Bhilwara Supergroup of Archean age in the Aravalli Craton, overlying Hindoli Group and underlying Sandmata Complex. Mangalwar Complex is a group of various gneisses and is divided into Lasaria Formation, Potla Formation, Suvana Formation and Kekri Formation. Lasaria Formation consists mainly of migmatites and composite gneisses with enclaves of tremolite-talc-chlorite schist, amphibolites, quartzite and crystalline limestone. Potla Formation consists of garnet-staurolite-biotite schist, kyanite-sillimanite-mica schist and felspathic biotite schist with subordinate quartzite, crystalline limestone, amphibolites and migmatitic gneisses. Suvana Formation covering a small area in the Potla Formation near Bhilwara, consists of migmatised amphibolites with minor bands of garnet-biotite-kyanite schist, calc-silicate rocks and quartzite. Potla and Suvana Formations constitute the bulk of metasediments of Mangalwar Complex. Kekri Formation consists of migmatites and composite gneisses, garnet-biotite schist and calc-silicate rocks which occurs as lenses within Potla Formation. Ranthambore Group and Rajmahal Formation/Quartzite are seen within Kekri Formation in the north. Ranthambore Group consists of series of quartzites and phyllites intruded with sills of dolerite as seen near Sawai Madhopur in Rajasthan. Rajmahal Quartzite/Formation consists of about 3 km wide massive quartzite with pebbly horizon at contact with mica schists of Kekri Formation. See Table 6.

M 24 Mangli Beds: Are equivalent to Panchet Formation of Gondwana Supergroup in Wardha Valley, Maharashtra.

M 25 Maniyara Fort Formation: An Oligocene age formation of Kutch, Gujarat, equivalent to Bermoti Series of Kutch. See Tables 20 and 38.

M 26 Manjir Formation: A Precambrian Formation of Neoproterozoic age of Higher Himalaya in Himachal Pradesh. See Precambrian of Himalaya.

M 27 Mar Formation: A Quaternary (Pleistocene to sub-Recent) age formation of Bikaner-Nagaur Basin in Western Rajasthan. See Table 16.

M 28 Maravattur Clay Formation: A formation of Uttatur Group of Trichinopoly area: See Trichinopoly area and Table 18.

M 29 Margo Formation: An Upper Triassic to Cretaceous age fossiliferous formation of Karakoram Range in Hunza and Laddakh areas of Jammu and Kashmir.

M 30 Marh Formation: A Paleocene to Eocene age formation of Bikaner-Nagaur Basin in western Rajasthan overlying the Palana Formation. See Table 16.

M 31 Marwar Craton: In the Aravalli Craton, overlying the Delhi Supergroup and its associated and intruded igneous and metamorphic bodies lies westward the Marwar Craton separated from the eastward Mewar Craton by Phulad Lineament. The Marwar Craton is of Neoproterozoic age and comprises of 3 units which from bottom to top are 1. Sirohi, Sindrath and Punagarh Groups; 2. Malani Igneous Suite and 3. Marwar Supergroup.

M 32 Marwar Supergroup: A sedimentary Basin of Marwar Craton in the Aravalli Craton in western Rajasthan which is about 2 km thick and ranges in age from Neoproterozoic to Eocambrian and consists of arenites, carbonates, evaporites and reddish argillites and well known potash bear in halite and gypsum deposits. The Supergroup overlies the Malani Igneous Suite, Sirohi Group and Aravalli and Delhi Supergroups. The Supergroup is deposited in two separate basins- larger the Nagaur Basin with many isolated sub-basins and smaller the Birmania Basin. Nagaur Basin consists of 3 Groups, which from bottom to top are Jodhpur Group, Bilara/Hansera Group and Nagaur Group. For Marwar Supergroup Permo-Carboniferous age has also been suggested by Das Gupta, 1998. See Tables 6 and 16 and see each Group separately.

M 33 Maspur Basalt Formation: See Abujhmar Group.

M 34 Mastoli Formation: A Precambrian formation of Higher Himalaya in Uttarakhand. See Precambrian of Higher Himalaya.

M 35 Matanomadh Formation: A Paleocene age formation of Kutch, Gujarat.

M 36 Mathaji ka Dungar Formation: An Eocene age formation of Barmer Basin in western Rajasthan, consisting of conglomerate and sandstone with casts of lamelli branchs, gastropods and leaf impressions which are commonly found. See Table 16.

M 37 Mayurbhanj Granite: A granitoid rock of the Singhbhum Craton occurring between The Singhbhum Granite and Dalma Volcanic Group. See Table 5.

M 38 Mayyanad Formation: The lowermost formation of Warkalli Group of Tertiary age in Kerala, consisting of sandstones, sand, clays and lignite. See Table 25.

M 39 Meekoceras zone: A thin Lower Triassic bed of Lilang Group, Spiti, Himachal Pradesh, consisting of only 1 m of fossiliferous thin bedded limestone and shale. Meekoceras and equivalent beds are also found in Byans and Painkhanda areas of Uttarakhand. See Table 13.

M 40 Megalodon Limestone/Para Stage: The lower bed of Upper Triassic age of Kioto Formation of Lilang Group, Spiti, consisting of 250 m of fossiliferous massive limestone and dolomite. See Table 13.

M 41 Meghalaya Group: Late Cretaceous to Early Paleocene age Group exposed near Therria in Meghalaya, NE India, comprising of Mahadeo and Um Sohryngkew Formations, 50 m and 250 m thick respectively.

M 42 Mesozoic Era: The Era extends in age from 261 Ma to 65.5 Ma and consists from bottom to top of Triassic, Jurassic and Cretaceous Periods.

M 43 Mesozoic succession of Kutch, Gujarat; Consists of Pachcham, Chari and Katrol Formations of Jurassic age and Umia Formation of Cretaceous age. See each formation and Table 15.

M 44 Mewar Gneiss: The Archean age Mewar Gneiss is the oldest member of the Aravalli Craton and forms basement to the Aravalli Supergroup and corresponds to the migmatites and gneisses of the Lasaria Formation of Mangalwar Complex. The Mewar Gneiss is composed of following 3 main components: 1. Basement TTG gneisses which are best exposed between Mavli and Pipalkhunt. The main rock type is the banded migmatitic biotite gneiss of TTG composition. 2. Basement granitoids intruding the TTG gneisses-e.g. Untala and Gingla granites. The Untala granite occurs near Untala (Vallabhnagar), near Udaipur and consists of diorite-tonalite-trondhjemite and grey to pink granite grading into granodiorite. Gingla granite occurs near Jaisamand (Dhebar) Lake near Udaipur. Gingla granite is partly magmatic and partly granitic. The migmatite has tonalite-trondhjemite composition, whereas the granite has granite-adamellite composition. 3. Basement inliers- these are granitoids occurring as inliers within the Aravalli fold belt e.g. the Sarara, Mando and Ahar River granite. Mewar Gneiss represents a part of earlier called Banded Gneissic Complex (BGC). See Table 6.

M 45 Middle Katrol: A bed of Katrol Formation in the Jurassic succession of Kutch, Gujarat. See Table 15.

M 46 Mikin Formation: Lowermost formation of Lilang Group of Spiti, Himachal Pradesh of Lower to Middle Triassic age. See Tables 13 and 14 for details of formations.

M 47 Mikir Hills Trap: Consists of volcanic rocks of Early Cretaceous to Permian age occurring in Mikir Hills and Dhansiri valley of Assam – Arakan Basin in NE India. See Table 23.

M 48 Milan Group: A Precambrian formation of Lesser Himalaya in Himachal Pradesh. See Precambrian of Himalaya.

M 49 Miliolite / Porbandar Stone: An Upper to Middle Pleistocene age formation of Gujarat, occurring all along the southern coast of Saurashtra from Gopinath in the east to Porbandar in the west. It is a favoured building stone composed of oolitic particles of calcareous sand and form a compact rock containing

Miliolite- a foraminifera. Miliolite is also called as Miliolite Formation. See Table 20.

M 50 Miliolite/Miliolite Limestone: A Pleistocene age rock of Kutch, Gujarat. The Miliolite Limestone occurs as small inliers within the rocky area as obstacle dunes, sheet deposits and valley fill deposits.

M 51 Mineralised Supracrustal Belts: These are mineralized supracrustal belts of Palaeoproterozoic age lying in the Aravalli Craton above the Mangalwar Complex and Hindoli Group and below the Aravalli Supergroup. There are many supracrustal belts of arenite-pelite-carbonate association with black shales and Banded Iron Formation containing rich deposits of lead-zinc minerals. A few important mineralized belts are Jahazpur Belt, Sawar Belt, Pur-Banera Belt, Rajpur-Dariba Belt, Rampur-Agucha Belt and Bhinder Belt. These mineralized belts have been correlated with the Zawar Belt of Udaipur Group of the Aravalli Supergroup. Conglomerate, quartzite, phyllite, dolomite, Banded Iron Formation, various types of schists, etc. form the common rock types of the mineral belts which host lead-zinc mineral deposits. Jahazpur Belt also includes Jahazpur Granite, Rampur-Agucha Belt hosts the largest lead-zinc deposits in the country and consists of varieties of gneisses, etc.

M 52 Miocene rocks of Karnataka: Miocene sediments are exposed along Kullur-Kavur road near Mangalore along the west coast. These rocks contain rich palynological assemblage consisting of angiosperm pollen and pteridophytic spores, much similar to that of Quilon and Warkalli beds.

M 53 Miocene rocks of Maharashtra: Consist of white and blue clays along with sandy and gravely material resting on Deccan Traps near Ratnagiri along the west coast in Konkan area. The clays are intercalated with thin carbonaceous seams of lignite containing fossilised fruits leaves along with mineral resin and pyrite.

M 54 Miri Formation: A Permian age formation of Lower Gondwana Supergroup in Arunachal Pradesh.

M 55 Mobile Belts: See Tectonic framework of India.

M 56 Mochia/Zawar Formation: See Udaipur Group.

M 57 Monotis Shales: An Upper Triassic age bed of Lilang Group, Spiti, consisting of 100 m of fossiliferous sandy and shaly limestone, shale and sandstone. See Table 13.

M 58 Moran Group: A Plio-Pleistocene age Group rich in palynoflora fossils, occurring in the Mikir Hills, Dhansiri valley and Upper Assam areas of NE India.

M 59 Morar Formation: See Gwalior Basin/Group.

M 60 Motur Formation: A Middle Permian formation of Gondwana Supergroup, equivalent to Barren Measures, occurring in the Satpura Basin and consisting of red clay, felspathic sandstone and grey shale with thin coal laminae. See Table 12.

M 61 Muddapur Dolomite: See Kaladgi Basin and Table 37.

M 62 Mulug Group: See Godavari Supergroup and Table 30.

M 63 Murree Formation: An Oligocene-Early Miocene age formation of Potwar-Jammu area in Jammu and Kashmir State. See Table 22.

M 64 Muschelkalk Lower: See Lower Muschelkalk.

M 65 Muschelkalk Upper: See Upper Muschelkalk.

M 66 Muth Quartzites/ Formation: A white quartzite formation of Devonian age occurring in the Tethyan Himalaya in Jammu and Kashmir, Himachal Pradesh (Spiti), and Uttarakhand (Byans & Malla Johar). In Jammu and Kashmir Muth Quartzites/Formation occurs in Warwan area, lower reaches of Sind valley, Shams Abari syncline of Handwar, Zanskar-Rupshu area, north of Penging Lake and Shylok valley, Naubagh valley and Saribal Range. See Table 11.

N 1 Naga Parbat Gneiss: A Precambrian Formation in the Higher Himalaya of Jammu and Kashmir. See Precambrian of Himalaya.

N 2 Nagaur Basin: The larger basin of Marwar Supergroup, consisting of 3 Groups, which from bottom to top are Jodhpur Group, Bilara/Hansera Group and Nagaur Group. See Marwar Supergroup.

N 3 Nagaur Group/Formation: The uppermost Group of Nagaur Basin, consisting of sandstones with basal conglomerate. The basal conglomerate contains assorted clasts of dolomite, chert, sandstone, feldspar and quartz. Red mottled sandstone interlayered with shale and siltstone overlies conglomerate. The top unit contains shale, marl, siltstone and pebbly sandstone. See Table 16.

N 4 Nagri Formation: See Siwalik Group and Table 28.

N 5 Nagum Formation: See Karewa Group and Table 24.

N 6 Nallamalai Group: This is the youngest group of the Cuddapah Supergroup. The Group is about 3500 to 6000 m. thick and consists of 5 formations, which from bottom to top are - 1. Nagari Quartzite consisting of quartzite, quartz wacke, shale and basal conglomerate; 2. Bairenkonda Quartzite, consisting of quartzite, quartz wacke and shale; 3. Pullampet formation, consisting of shale, felsic tuff, barytes, dolomite and quartzite; 4. Cumbum Formation, consisting of slate, phyllite, chert and dolomite; 5. Srisailam Quartzite, consisting of glauconitic ferruginous quartzite with shale. See Cuddapah Supergroup and Table 7.

N 7 Nambi Formation: Lowermost formation of Usur Group of Godavari Supergroup. See Godavari Supergroup and Table 30.

N 8 Namsang Formation: An Olicene age formation of Upper Assam Valley and Naga Hills Fold Belt area of Assam-Arakan Basin in NE India. The formation is equivalent to Dupitila Formation. See Table 23.

N 9 Nandgaon Group: A Proterozoic age Group of Koteri-Dongargarh Orogen of Bastar Craton, consisting from bottom to top of Dargah Formation, Bijli Rhyolite and Pitepani Basalt. See Table 4.

N 10 Nareda Formation: An Early Eocene age formation of Kutch, Gujarat, lying below the Harudi Formation.

N 11 Narha Beds: A small horizon of sandstone and shale near Narha village in Kutch, Gujarat, equivalent to Umia Formation of Cretaceous age which is considered to be of Gondwana Supergroup. See Tables 12 and 15.

N 12 Nari Formation: An Oligocene age formation of Sindh-Salt Range area of Pakistan.

N 13 Narmada Formation: A Holocene age formation of Gujarat Mainland. See Table 20.

N 14 Narpuh Sandstone Member: A member of Sylhet Sub-group of Jaintia Group of Eocene age in the Khasi and Jaitia Hills area of NE India. See Table 23.

N 15 Neil West Coast Formation: A Pliocene to Pleistocene age formation of Andaman-Nicobar Islands. See Table 21.

N 16 Nepheline Syenite of Kishangarh: A suite of Alkaline rocks with rarely found nepheline syenite as dominant component, occurring around Kishangarh, near Ajmer in Rajasthan. The Nepheline Syenite body is about 25 km long and 1.0 to 1.5 km wide. At the eastern contact of this body occurs the Sandmata Complex of the Bhilwara Group, while on the western contact occurs pre-Delhi supracrystals consisting of quartzite, mica schist, marble and amphibolites, locally called Kishangarh Group. The Nepheline Syenite consists of 20 to 30 percent of felspathoids and variable amounts of mafic minerals and principal feldspar minerals. Alkaline mafic rocks like felspathic ijolite as lenticular patches, theralite as xenoliths and comptonite as dykes occur locally within the Syenite. Complex folding and gneissic foliation at the margins of the Syenite body suggests that it is a synkinematic intrusion associated with Delhi Orogeny. See Table 8.

N 17 Nerinea Beds: A term used earlier to identify beds bearing fossils of Nerinea (a gastropod), of the Niniyur Formation of Cretaceous age of the Trichinopoly area in Tamil Nadu.

N 18 Newania Carbonatite: It is the smallest carbonatite body found near Udaipur, intruded in to the Untala Granite of Archean age in the Aravalli Craton.

N 19 Newer Dolerite: A group of basic dykes traversing the entire Singhbhum Craton and consisting predominantly of quartz-dolerites, peridotites, norites, lamphrophyres, microgranites and leucogranophyres. See Table 5.

N 20 Neyveli Lignite Field: It is India's most important lignite field, existing in the Thanjavur district of Tamil Nadu. The lignite deposits which are of Miocene age, occur in the Cuddalore Sandstone Formation. The lignite field covers an area of about 40 x 10 sq.km. and the lignite seams are 8 to 22 m thick and are underlain by sands with artesian aquifers which exert an upward pressure.

N 21 Nimar Sandstone: Lowermost member of Bagh Bed Formation. See Bagh Beds.

N 22 Nimoloksa Group: A Group of Lilang Supergroup of Triassic age in Spiti area of Himachal Pradesh. See Lilang Supergroup and Table 14.

N 23 Niniyur Group/Formation: An Early Paleocene age Group/Formation of Trichinopoly area in Tamil Nadu. See Trichinopoly area and Table 18.

N 24 Niralkeri Breccia: See Kaladgi Basin and Table 37.

N 25 Nishatbagh Bed: A Lower Permian age formation of Kashmir area with plant fossils e.g. of Gangamopteris, having affinity with Gondwana Supergroup and consisting of black shale/slate, siltstone and bands of arenite.

N 26 Niti Limestone Member: A member of Mikin Formation of Triassic age Lilang Supergroup, Spiti. See Lilang Supergroup and Table 14.

N 27 Niti Member: A member of Shal Shal Formation of Triassic age in Painkhanda area of Uttarakhand, consisting of massive nodular limestone. See Table 26.

N 28 Nodular Limestone, Bagh: A Cretaceous (Albian) age bed of Bagh Beds. See Bagh Beds.

N 29 Nodular Limestone, Spiti: A Middle Triassic age bed of Lilang Group, Spiti, consisting of 20m of fossiliferous hard nodular limestone. See Lilang Group and Table 13.

N 30 Nunuluka Formation: An Upper Triassic age formation of Spiti area in Himachal Pradesh, consisting of Quartzite Series. A series of Up. Trissic age, overlying the Monotis Shales in the Lilang Group, Spiti. See Tables 13.

O 1 Offshore Basins of India: Need for petroleum products (oil, gas, etc.) induced study of offshore basins. The major offshore basins are : 1. Bengal Basin; 2. Mahanadi Basin; 3. Krishna - Godavari Basin; 4. Pennar Basin; 5. Palar Basin; 6. Cauvery Basin; 7. Konkan-Kerala Shelf; 8. Bombay Offshore Basin; 9. Kutch and Saurashtra Basin and 10. Andaman-Nicobar Basin.

O 2 Older Metamorphic Group (OMG): OMG forms one of the oldest rocks of the Singhbhum Craton and occurs as a series of enclaves in Champua, Shankarpur, Bahalda, etc. areas. The main rock types of OMG are biotite-muscovite schist and varities of quartzites and amphibolites. See Singhbhum Craton and Table 5.

O 3 Older Metamorphic Group of NE India: It is enclave suite of ancient supracrustal rocks best exposed in the western part of Khasi and Jaintia Hills, which is an extension of Chhotanagpur Gneiss of the Satpura Mobile Belt in NE India. The enclave suite consists of cordierite-biotite (+/- sillimanite) gneiss, quartz-sillimanite schist, Mg-Al pelites, Calc-silicate rocks, Banded Iron Formation, Amphibolites and pyroxene granulites, Charnockites and Biotite Gneiss.

O 4 Older Metamorphic Tonalite Gneiss (OMTG): OMTG forms one of the oldest rocks of the Singhbhum Craton and is predominantly of tonalite-trondhjemite-granodiorite (TTG) composition with subordinate granodiorite, quartz diorite and monzodiorite. The gneisses are well foliated and locally banded and contain plagioclase, quartz, biotite and subordinate K-feldspar with accessory zircon, sphene and apatite. See Singhbhum Craton and Table 5.

O 5 Ongarbira Volcanics: A volcanic group of the Singhbhum Craton occurring in association with the Koira Group, lying over the Dhanjori Group and below the Singhbhum Group. The Volcanis consists of low- K tholeiites, ultramafic flowfs, tuffs, cherty felsic lavas and gabbro-pyroxenite intrusive with prominent pillow structures. See Singhbhum Craton and Table 5.

O 6 Ophiceras Zone: A Lower Triassic bed of Lilang Group, Spiti, consisting of 3m of fossiliferous grey limestone. See Triassic of Spiti and Table 13.

O 7 Ordovician-Silurian rocks: Occurs together and it is practically not possible to reliably diffentiate between Ordovician and Silurian rocks which consists of volcanic associated with slate, slates and phyllites in Kashmir; conglomerate, quartzite, siltstone and shale in Spiti-Kinnaur area of Himachal Pradesh and shales, sandy limestone and crinoidal breccias in Kumaon-Garhwal areas of Uttarakhand. Fossils are reported from Ordovician-Silurian rocks of Shans Abari, Kashmir near Handwar in Jammu & Kashmir.

O 8 Otoceras Zone: The lowermost Triassic bed of Lilang Group, Spiti, consisting of 6m of fossiliferous brown limestone. See lilang Group and Table 13.

O 9 Ottakovil Formation: A Cretaceous age formation of Ariyalur Group of Trichinopoly area. See Table 18.

P 1 Pachcham basal beds: The lowermost Jurassic Beds of Pachcham Formation in the Mesozoic succession of Kutch, Gujarat. These beds are also known as Kuar Bet Beds. See Mesozoic succession of Kutch, Gujarat and Table 15.

P 2 Pachcham Coral Bed: A Jurassic age Bed of Pachcham Formation in the Mesozoic succession of Kutch, Gujarat. See Mesozoic succession of Kutch, Gujarat and Table 15.

P 3 Pachcham Formation: A Jurassic age formation in the Mesozoic succession of Kutch, Gujarat. See Mesozoic succession of Kutch, Gujarat and Table 15. Due to affinity with Gondwana Supergroup the formation is also shown in Table 12.

P 4 Pachcham Shell Limestone: A bed of Pachcham Formation of Jurassic age in the Mesozoic succession of Kutch, Gujarat. See Mesozoic of Kutch, Gujarat and Table 15.

P 5 Pachmarhi Formation: A Gondwana Supergroup formation of Lower Triassic age, equivalent to Panchet Formation, occurring in the Mahadeo hills of Satpura Basin and consisting of sandstone with pebbles. See Table 12.

P 6 Pairi Group: The sediments of Khariar Basin are grouped as Pairi Group. See Table 35 and Khariar Basin.

P 7 Pala Lahara Gneiss: Occurs in the southwestern part of the Singhbhum granite in the Singhbhum Craton and consists of dark bands of hornblende and migmatite alternating with white quartzo-felspathic bands. See Table 5.

P 8 Palana Formation: A Paleocene-Eocene age subsurface formation of Bikaner-Nagaur Basin in western Rajasthan, containing lignite deposits. See Table 16.

P 9 Palaram Shale Formation: A Proterozoic age formation of Mulug Group of Godavari Supergroup in the Pakhal/Mallampalli Belt in the Pranhita-Godavari Basin. See Godavari Supergroup and Table 30.

P 10 Paleozoic Era: This is the oldest era of Phanerozoic Eon, overlying the Precambrian Proterozoic Eon. The Paleozoic Era ranges in age from 542 to

261 Ma and consists from bottom to top of Cambrian, Ordovician, Silurian, Devonian, Carboniferous and Permian Periods. See Table 1.

P 11 Paleozoics of Kashmir: The Paleozoic succession of Kashmir is exposed in the Lidder Valley and includes rocks of Silurian, Devonian and Permian age. See Table 11.

P 12 Pali Formation: A Gondwana Supergroup formation of Middle Permian to Lower Triassic age in the Rewa area of Son-Mahanadi Basin, consisting of sandstone, pebble beds and red clays. The formation is divisible into Lower, Middle and Upper Members. See Table 12.

P 13 Panchet Formation: A Lower Triassic age formation of Gondwana Supergroup, consisting of sandstone, shale and siltstone in its type area in Raniganj basin. See Table 12.

P 14 Panchet Group: A group of Lower Gondwana Supergroup, comprising of Panchet and Supra-Panchet Formations in the Damodar valley coal basin. See Table 12.

P 15 Pandyan Mobile Belt: This Proterozoic age Mobile Belt lies near the southern tip of Indian Peninsula, south of Dharwar Craton separated by Palghat-Cauvery Shear Zone. The Belt is divided into 3 major tectonic units- Marginal Zone, Madurai Block and Trivendrum Block. The Marginal Zone predominantly consists of gneisses, charnockites and supracrustal enclaves of Dharwar Craton. Madurai Block or the Central Zone consists essentially of quartzites, Mg – Al pelites, garnet-sillimanite gneisses and calc-silicate rocks of Khondalite Group. The Trivendrum Block or the Southern Zone consists of Ponmudi (Keral Khondalite Belt) sub-Block and Nagercoil sub-Block. The Ponmudi sub-Block consists of garnet-sillimanite = graphite gneisses, garnet-cordierite gneisses, calc-silicate rocks and marbles, quartzites, pyroxene granulites, garnet-biotite gneisses and garnet bearing quartzo felspathic gneisses, incipient charnokites and massive and banded charnockites. The Nagercoil sub-Block consists mainly of charnockites and gneisses containing several layred anorthositic to noritic bodies.

P 16 Panjal Volcanics: A Lower Permian age formation overlying the Nishatbagh Formation of Kashmir sub-basin and consisting of basic volcanic rocks, e.g. basalt, andesitic basalt and a fefw intermediate and acidic rocks. See Table 11.

P 17 Papaghni Group: This is the oldest group of the Cuddapah Supergroup with thickness of about 2100 m. The Group consists of two formations - older the Gulcheru Quartzite which consists of Basal Conglomerate, arkose, quartzite with shale interbeds. The younger formation is Vempalle Formation with Kuppalapalle Volcanics, which consists of stromatolitic dolomite, chert, quartzite and basic sills and flows. See Table 7

P 18 Papaghni Sub-basin: It is a sub-basin of Cuddapah Basin of Dhaewar Craton, which is one of the biggest Purana Basins of Paleoproterozoic age. The papaghni sub-basin houses the Papaghni Group of the Cuddapah Supergroup. See Papaghni Group and Table 32.

P 19 Par Formation: See Gwalior Basin/Group.

P 20 Para Stage: See Megalodon Limestone and Tables 13 and 14.

P 21 Parahio Series: A Cambrian age formation of Spiti area in Himachal Pradesh, associated with the Kunzun La Formation, and consisting of fossiliferous dolomite bands, etc.

P 22 Parh Formation: A subsurface Cretaceous age formation of Jaisalmer Basin, Rajasthan encountered during drilling, consisting predominantly of greenish grey argillaceous limestone and marl. See Table 16.

P 23 Pariwar Formation: A Cretaceous age formation of Jaisalmer Basin, Rajasthan, consisting of fossiliferous sandstone. See Table 16.

P 24 Parsoi Formation: A Proterozoic age formation of Mahakoshal Group of the Satpura Mobile Belt, consisting of greywacke-argillite, carbonaceous phyllite, felspathic quartzite and basal polymict conglomerate.

P 25 Parsora Formation: A Gondwana Supergroup formation of Upper Triassic age in the Rewa area of Son-Mahanadi Basin, consisting of purple white kaolinitic sandstone. See Table 12.

P 26 Pashkyum Formation: An Eocene age formation of Kargil area in Kashmir, consisting of shales, sandstones and conglomerates. See Eocene of Kashmir.

P 27 Passage Beds: Beds of Carboniferous age lying between Fenestella Shale and Syringothyris Limestone in the Palaeozoic succession of Lidder Valley, Kashmir. In the revised classification these beds are equated with Aishmuqam Formation. See Aishmuqam Formation and Table 11.

P 28 Pattipalli Quartzite Formation: A Proterozoic age formation of Mulug Group in Pakhal/Mallaplli Belt of the Godavari Supergroup. See Table 30.

P 29 Penganga Group: A Proterozoic age group of Godavari Supergroup in Pakhal/Mallampalli Belt. See Table 30.

P 30 Peninsular Gneiss: Also called Grey Gneisses, are more than 3000 million years old and occurs in the western Dharwar Craton, over lying the Sargur Group and under lying the Dharwar Supergroup. The Peninsular Gneiss consists of polyphase migmatites i.e. multiple migmatites of TTG (Tonalite-Trondhjemite-Granodiorite) composition, and homogenous granitoids i.e. trondhjemite to granodiorite plutonic intrusions into the migmatites. See Table 2.

P 31 Permian rocks: Bomte Member of Bichom Formation of Arunachal Pradesh containing Lower Prmian fauna and Bahreli Formation containing Gangamopteris flora. Abor Volcanic Formation and associated Inter-trappeans have yielded Lower Permian micro flora. Yamuna Formation contain marine invertebrate fauna of Upper Permian age. In Assam Permian rocks are scarce and localised at Singrimari in Golapara district where Miri Formation of Lower to Middle Palaeozoic age is exposed. In Byans and Painkhanda of Uttarakhand a sequence over Muth Quartzites occurs in Niti, Malla Johar and Byans containing Early Permian fauna. Rocks of Kuling Shale Formation in Niti Pass Hotigad and Raulibagar sections in Painkhanda contain palynoflora of Permian age, equivalent to Raniganj Formation of Gondwana Supergroup. In Doggada and adjoining area of Garhwal Boulder Slate are similar to upper part of Agglomeratic Slate of Permian age. Other Permian formations are Panjal Volcanics and upper part of Agglomeratic Slate, Karharbari, Barakar, Barren Measures and Raniganj Formation and their equivalents.

P 32 Phulad Lineament: It divides the Aravalli Craton into Mewar Craton on the east and Marwar Craton on the west and also marks the western boundary of Delhi Fold Belt.

P 33 Phulad Ophialite Suite: A shear zone of the South Delhi Fold Belt in the Aravalli Craton exposed intermittently for a distance of 500 km from Khedbrahma through Kui in Gujarat to Ajmer in Rajasthan, separating the Bhim Group, Sendra Group and Barotiya Group near Kamblighat. The shear zone which is 40 km wide consists of epidiorite, amphibolites, pyroxene granulite, metagabbro and ultramafic rocks.

P 34 Pindobal Formation: A Carboniferous to Lower Permian age formation of Lidder Valley in Kashmir. See Table 11.

P 35 Pinjor Formation: A Pleistocene age formation of Siwalik Group. See Table 28.

P 36 Piram Beds: Upper Miocene to Lower Pliocene age fossiliferous beds consisting of conglomerate, silty sandstone and clay in the Ghoga area, well developed in the offshore island of Piram in Saurashtra in the Gulf of Cambay, Gujarat, overlying the Gaj Formation. See Table 20.

P 37 Pleistocene sediments: Areas of main Pleistocene sediments of India are; a. Kurnool, Ghod, Manjara, Pravara and Godavari; b.Central Narmada (Upper group) Palmar; c. Central Narmada (Lower group); and d. Lower Karewa, Pinjor of Upper Siwaliks.

P 38 Po Formation: A Carboniferous age formation of Kanawar Group, consisting of shale, siltstone, sandstone/quartzite, overlying the Lipak Formation in the Spiti-Kinnaur, Himachal Pradesh and Kumau, Uttarakhand sub-basins. In Kinnaur-Spiti sub-basin the Po Formation is overlain by Ganmachidam Formation.

P 39 Pokran Boulder Bed: See Jodhpur Group:

P 40 Porbandar Stone: See Miliolite:

P 41 Porlob Group: An Upper Cretaceous age Group of Andaman Nicobar Islands, consisting of porcellaneous limestones or marbles, cherts, jasper and quartzites. On weathering these rocks at splaces give rise to bright red lateritic soil. See Cretaceous of Andaman and Nicobar Islands.

P 42 Port Blair Formation: An Upper Cretaceous-Palaeogene age formation of Andaman-Nicobar Islands. See Table 21.

P 43 Port -Meadow Formation: Oldest formation of Cretaceous-Palaeogene age in the stratigraphic succession of Andaman-Nicobar Islands. See Table 21.

P 44 Potla Formation: See Mangalwar Complex and Table 6.

P 45 Prang Member/Formation: A Paleocene to Early Eocene age Member/ Formation of Sylhet Sub-group in the Khasi and Jaintia Hills area of NE India. See Table 23.

P 46 Pranhita Sandstone Formation: A Formation of Penganga Group of Godavari Supergroup of Proterozoic age. See Godavari Supergroup and Table 30.

P 47 Pranhita-Godavari Basin: This is one of the Purana or Proterozoic sedimentary Basins of Peninsular India, situated at the junction of Dharwar and Bastar Cratons. The Basin exists in two parallel belts or sub-basins, the western one is called the Pakhal or Mallampalli Belt and the eastern one is called Albaka Belt. The Basin houses the Godavari Supergroup. See Godavari Supergroup and Tables 30 and 32.

P 48 Pranhita-Godavari Valley: It is a NW-SE trending valley in Andhra Pradesh, named after the Pranhita and Godavari rivers of the area, and extends from Elur at its SE end to Asifabad on its NW end, with approximate length of 350 km and width of 60 km. The valley is occupied with the fossiliferous Gondwana Supergroup sediments as shown in the Table below:

Age	**Gondwana Supergroup**	**Formation**	**Lithology**
Lr. Cretaceous ?	Upper Division	Chikiala	Sandstones and conglomerates
Lr. Cretaceous		Gangapur	Sandstones and mudstones with interbedded sandstones and concretions
Jurassic		Kota	Sandstones, siltstones and mudstones with limestone bands
Triassic		Dharmaram	Sandstones and mudstones
		Maleri	Mudstones, sandstones and calcirudites
		Bhimaram	Sandstones and mudstones
	Lower Division	Yerrapalli	Mudstones and calcirudites
Permian to Triassic		Kamthi	Sandstones and siltstones

P 49 Precambrian of Himalaya: Precambrian rocks in Himalaya occur in the Lesser Himalaya and in the Higher Himalaya. Sequence from north to south is as follows:

Tethyan Zone

-------- Trans Himadri Thrust --------

-------- Higher Himalaya --------

-------- Main Central Thrust --------

-------- Lesser Himalaya --------

-------- Main Boundary Thrust --------

-------- Siwalik Hills --------

-------- Main Boundary Fault --------

-------- Indo-Gangetic Plain --------

The tectonic-stratigraphy of Himalaya consists of eleven sedimentary cycles, of which first four cycles pertain to Precambrian age. Precambrian in Higher and Lesser Himalaya occur in Jammu and Kashmir, Himachal Pradesh, Uttarakhand, Nepal, Sikkim-Darjeeling, Bhutan and Arunachal Pradesh. Nepal and Bhutan being outside India are not mentioned here. The Precambrian of Himalaya in India are as follows: (1) **Jammu and Kashmir:** HH (Higher Himalaya)-C I (Cycle one)-Naga Parbat Gneiss, C II Salkhala Gr (Group) with Tanwal Formation, C III & IV - Karihul Formation, Lolab Formation; LH (Lesser Himalaya) CII Befliaz Formation, Kalamund Formation, C III - Ramban Formation (Sona Volcanics), Gamir and Baila Formations, C IV - Ramsu, Macchal, Bhimdasa and Eo (Eocambrian) Sincha Formations. (2) **Himachal Pradesh**: HH C I –Vaikrita Gr, Suru, Giambal, Rohtang, TSO Morari Gneiss, CII- Chamba Formation, Salkhala Gr (with Tanwal Formation), Jutogh & Kullu Grs; C III & IV- Kunzun La, Batal, Bhaderah and Manjir Formations; LH-C I- Wangtu, Bandal, Almora, Askot, Baijnath and Dharamgarh Gneisses; C II- Rampur Gr (Banjar Volcanics), Sundergarh Gr (Mandi Volcanics), Largi Gr, Shali Gr; C III- Simla Gr (Tatapani volcanics), Jaunsar Gr (Peontra Volcanics); C IV- Blaini Boulder Bed Formation/Gr, Krol Gr; Eo-Tal Gr, Milan Gr; (3) **Uttarakhand:** HH-C I-Vaikrita Gr (Central Crystallines), C II- Haimanta Gr, C III & IV- Dudatoli & Martoli Formations; LH-C I- Annapurna Gneiss; C II - Garhwal Gr (Rautgara, Tejam, Berinag, Pithoragarh, Naraul and Bhowali Grs); C III Dirang Gr; (4) **Sikkim-Darjeeling:** HH- C I- Kanchanjhanga Gneiss; LH- C I - Thimphu Gr; C II- Daling Gr; C III – Gorubathan Formation, Baxa Gr. (5) **Arunachal Pradesh:** HH- C I – Se La Gr; C III & IV - Lum La Formation; LH – C II – Bomdila Gr; C III – Dirang Gr; C IV – Dedza Formation.

P 50 Productus Limestone: A Permian age limestone formation of Salt Range area in Pakistan.

P 51 Productus Shale: A Permian age formation, which in Jammu and Kashmir forms part of Zewan Formation, in Spiti area in Himachal Pradesh Productus Shale is represented by Salooni Formation and in Uttarakhand by Boulder Slate of Doggada and adjoining areas of Garhwal.

P 52 Proterozoic Sedimentary Basins: See Purana Basins.

P 53 Punagarh Group: The Group overlies the Sindreth Group of the Marwar Craton of the Aravalli Craton. Punagarh Group consists of shale, phyllite, mica schist and quartzite along with a few flows of pillowed meta basalt, amygdular basalt, volcanic breccias tuff and bedded chert. See Table 6.

P 54 Purana Basins: These are Proterozoic age sedimentary basins of Indian peninsula, deposited over the Precambrian crystalline basement consisting of gneisses, granites and schists. These basins are nearly flat, partly deformed and consists predominantly of orthoquartzite-shale-carbonate suites. Purana basins are rich in limestone and dolomite and other economic minerals like bedded barites, phosphorite, chrysolite asbestose, base metals and building stones. For details see Tectonic framework of India and Table 32.

P 55 Putnur Limestone Formation: A formation of Pakhal/Mallampalli Belt of Godavari Supergroup. See Table 30.

Q 1 Qazil langer Formation: A Cretaceous-Paleocene age fossiliferous formation of Karakoram Range in Hunza and Laddakh areas of Jammu and Kashmir State. The formation consists of red conglomerate with interbedded carbonaceous shale, sandstone and thin bedded limestone. See Cretaceous of Jammu and Kashmir.

Q 2 Quartzite Series: An Upper Triassic age bed of Lilang Group, Spiti, consisting of 100 m of fossiliferous quartzite with grey limestone and black shale. See Table 13.

Q 3 Quaternary sediments of Andhra Pradesh: These are well developed along the east coast, especially near mouths of Godavari, Krishna and Penner rivers and many other smaller rivers discharging into Bay of Bengal. Fossiliferous horizons occur in Gundlakamma and munneru river basins and in the flood plain of river Pampa.

Q 4 Quaternary sediments of Central India: These include Pleistocene and Holocene sediments of Belan and Narmada valleys of western Maharashtra including Godavari, Manjra, Krishna, Ghod etc. in central India which are richly fossiliferous and consists of gravel, sand, silt, clay, conglomerate, sandstone, etc. The Holocene sediments are essentially terrace deposits.

Q 5 Quaternary sediments of Gujarat Mainland: These include Late Pleistocene age sediments of Narmada, Mahi and Sabarmati rivers. The sediments are mainly marine bluish clay, overlain by alluvial fan and alluvial plain sediments. Also present are Middle-Late Holocene sediments as valley fill terraces. These Quaternary sediments contain fossils of foraminifera.

Q 6 Quaternary sediments of Himachal Pradesh: The Quaternary sediments of Himachal Pradesh are as shown below:

Age		**Lithology**
Holocene		Alluvium and low terraces
Up. Pleistocene		Kangra, Palampur and Bilaspur terraces. Relics of earlier major glaciations
Mid. Pleistocene	Upper Siwalik rocks	Neogal Beds and Boulder Conglomerate
Lr. Pleistocene and Up. Pliocene		Mainly conglomerate with minor sandstone and claystone

Q 7 Quaternary sediments of Kali Valley, Kumaon, Uttarakhand: See Table 27.

Q 8 Quaternary sediments of Maharashtra: These consists of alluvial deposits of Tapi, Purna, Godavari, Wardha, Penganga and Wainganga rivers and contain a few fossils in certain places. Besides alluvial deposits, there are Beach rocks of Middle to Late Holocene age which occur as isolated patches along the coast of Arabian sea and comprise conglomerates, shell limestone bands and consolidated dune sands called "Karal'.

Q 9 Quaternary sediments of Tamil Nadu: These occur mainly along the east coast and along some major river valleys. Coromandal Formation of Holocene age is reported to occur below the coastal beach sand in the northern coast of Chennai and Tiruvallur districts. Red sands called 'Teri' of Quaternary age are found in the Tuticorin district, covering gritty sandstones and shell limestone. Vertebrate fossils are found in calcareous gritty sandstone at Soyamali in Tirunelveli district.

Q 10 Quaternary sub-Era: The term Quaternary does not find mention in the International Stratigraphy Chart, 2004 issued by the International Commission on Stratigraphy. The term is however still in use and is considered as a sub-era of Cenozoic Era. See Table 19.

R 1 Rabanpalli Formation: See Bhima Basin/Group.

R 2 Raghvapuram Formation: A Gondwana Supergroup formation of Lower Cretaceous age, occurring in the Krishna-Godavari coastal area of Andhra Pradesh and consisting of shale, siltstone and micaceous sandstone with plant and animal fossils. See Table 12.

R 3 Raialo Group: It is the oldest formation of Lalsot-Bayana and Alwar Basins of North Delhi Fold Belt of the Delhi Supergroup. In the Lalsot-Bayana Basin the Raialo Group consists of basal conglomerates and pebbly quartzites, overlain by volcanics. In the Alwar Basin the Raialo Group consists of carbonate sediments including the famous Makrana Marble, followed by volcanics. See Table 8.

R 4 Rajahmundry Formation: An Eocene to Miocene age formation occurring in East and West Godavari districts of Andhra Pradesh. The formation overlies the Deccan Traps and underlies laterite covering. The lowest member of the formation is Rajahmundry conglomerate and sandstone, over which lies Bommuru claystone and Pidungoya sandstone, also carbonaceous clay and lignite and finally lies the Dowleswaram sandstone at the top.

R 5 Rajgarh Group: It is the lower group of Bhim Basin of South Delhi Fold Belt of the Delhi Supergroup and consists of cross-bedded quartzite, pelitic schist and calc-gneiss metamorphosed in upper amphibolites facies. According to some Rajgarh Group is of pre-Delhi age when it is correlated with the supracrustals surrounding Anasagar gneiss.

R 6 Rajmahal Basin: Exists in the NE part of Jharkhand State and includes the Rajmahal Hills and contains Talchir, Barakar and Dubrajpur Formations of the Gondwana Supergroup, which are overlain by Rajmahal Formation, consisting of Rajmahal Traps and associated inter-trappean beds.

R 7 Rajmahal Formation: A fossiliferous formation of Lower Cretaceous age of Gondwana Supergroup occurring in the Rajmahal Basin of Jharkhand and West Bengal and consisting of Rajmahal Traps (basaltic lava flows, etc.) and associated inter-trappean beds. See Table 12.

R 8 Rajmahal Inter-trappeans: The inter-trappean beds are associated with the Rajmahal Traps.

R 9 Rajmahal Traps: Are of Early Cretaceous age and consists of volcanic basaltic lava flows, also called Rajmahal Basalts/Volcanics. The Traps are associated with intertrappean beds of early Cretaceous age.

R 10 Rama Formation: A formation of Upper Triassic age of Lilang Supergroup, Spiti. See Table 14.

R 11 Ramagiri Formation: A formation of Sullavai Group of Pakhal/Mallampalli Belt of the Godavari Supergroup of Proterozoic age. See Godavari Supergroup and Table 30.

R 12 Ramania Stage: The lower stage of Bermoti Series of Oligocene age, in Kutch, Gujarat, also encountered in Lakhpat-I well, Kutch, Gujarat. See Tables 20 and 38.

R 13 Rambakot Member: Lowest Member of the Shal Shal Formation in the Triassic succession of Niti area, Painkhanda, Uttarakhand, consisting of limestone and shale. See Table 26.

R 14 Ramban Formation: A Precambrian formation of Lesser Himalaya in Jammu and Kashmir. See Precambrian of Himalaya.

R 15 Ramnad Beds: An exposure of Upper Gondwana Beds at Ramnad, near Chennai, consisting of boulder beds, conglomerate and micaceous sandstone containing plant fossils. Ramnad Beds are probably equivalent to or part of Satyavedu Formation.

R 16 Rampur Group: A Precambrian group of Lesser Himalaya in Himachal Pradesh. The Group includes Banjar Volcanics. See Precambrian of Himalaya.

R 17 Ramsu Formation: A Precambrian formation of Lesser Himalaya in Jammu and Kashmir. See Precambrian of Himalaya.

R 18 Randha Formation: A Proterozoic age formation of Barmer Basin in Western Rajasthan, lying above the Malani Igneous Suite and below the Birmania Formation. See Table 16. R 19 Rangamal Formation: A Cambrian to Middle Silurian age formation of Hapatnar Group in the Palaeozoic succession of Lidder Valley in Kashmir. See Table 11.

R 20 Rangapahar Formation: A Pliocene age formation, equivalent to Dupitila Formation, occurring in the Mikir Hills and Dhansiri Valley and Naga Hills Fold Belt areas of Assam-Arakan Basin in NE India. See Table 23.

R 21 Rangapani Formation: A formation of Pliocene to Pleistocene age and equivalent to the lower part of Dihing Formation, occurring in the Garo Hills area of Assam-Arakan Basin in NE India. See Table 23.

R 22 Rangrik Formation: An upper Triassic age formation of Nimoloksa Group of the Lilang Supergroup, Spiti. See Table 14.

R 23 Raniganj Formation: An Upper Permian age formation of Gondwana Supergroup, overlying the Barren Measures. The formation consists of sandstones, shales and workable coal seams as in the Raniganj Coal field. Lower Kamthi, middle Pali and Bijori are its equivalent formations. See Table 12.

R 24 Ranikot Formation: A Paleocene age formation of Sindh-Salt range area of Pakistan. See Table 22

R 25 Ranipet Formation: See Dharangadhara Group.

R 26 Ranthambore Group: See Mangalwar Complex and Table 6.

R 27 Ras Marble: An associate group of South Delhi Fold Belt of the Delhi Supergroup, previously regarded as a part of the Raialo Group, is now regarded as part of Delhi Supergroup with Barr Conglomerate at the base. Ras Marble occurs in the strike continuity of Makaran Marble and is bounded by granites and migmatites that are believed to be of pre-Delhi age.

R 28 Ratanpur Member: Lowermost member consisting of clay of Oligocene age of Gaj Formation of Saurashtra area in Gujarat.

R 29 Recent: The term does not find mention in the International Stratigraphic Chart, 2004, issued by international Commission on Stratigraphy. However the term is used to indicate the latest Period/Epoch just younger to Pleistocene Epoch and equivalent to Holocene Epoch.

R 30 Rehmanni Beds: Middle Callovian age beds of Chari Series in the Jurassic succession of Kutch, Gujarat. See Table 15.

R 31 Renji Formation: It is equivalent to Barail Group of Oligocene age, consisting of sandstone and occurring in Khasi-Jaintia Hills, Mikir Hills, Dhansiri Valley, Surma Valley and Tripura Fold Belt areas of Assam-Arakan Basin in NE India. See Tables 22 and 23.

R 32 Rewa Group: This is a group of Vindhyan Supergroup. For details see Table 9.

R 33 Rewak Formation: Uppermost formation of Eocene age of Jaintia Group in the Garo Hills area of NE India. See Table 23.

R 34 Rawali Bagar Group: See Malla Johar Supergroup and Table 17.

R 35 Rikhabdeo/Rishabhdeo ultramafics: Occurs with the Lunavada Group of the Aravalli Supergroup and lies over the Jharol Group and consists of serpentinites and talc-carbonate rocks. See Table 6.

R 36 Rimkhim Member: A member of Shal Shal Formation in the Triassic succession of Niti area, Painkhanda, Uttarakhand, consisting of limestone and shale. See Table 26.

R 37 Rishkobal Formation: A formation of Hapatnar Group of Cambrian to Middle Silurian age in the Palaeozoic succession of Lidder Valley, Kashmir.

R 38 Rohtang Gneiss: A Precambrian gneiss of Higher Himalaya in Himachal Pradesh. See Precambrian of Himalaya.

R 39 Rongtong Formation: An Upper Triassic age formation of Sanglung Group of Lilang Supergroup, Spiti. See Table 14.

R 40 Round Formation: A formation of Early to Middle Miocene age, of Archipelago Group in the Andman Nicobar Islands. The formation is also called Strait Formation. See Table 21.

S 1 Sabari Basin/Group: It is one of the Puran or Proterozoic Sedimetary Basins of Meso-Neoproterozoic age of Bastar Craton and is separated from the Indravati Basin by the Darba-Jhiram ghat occupied by granites and Bengpal Group and has faulted contact in the north with Indravati basin. Chedrapal outlier which consists of conglomerate, quartzite and some shale connects the Indravati and Sabri basins. The Sabri basin is occupied by the Sabri Group which consists of a basal conglomerate and orthoquartzite, overlain by limestone, and followed by shale. See Table 32.

S 2 Sakoli Group: is of Proterozoic age and is considered equivalent to Chilpi Group of Bastar Craton and comprises from bottom to top of Gaikhuri, Dhabetekri, Bhivapur and Pawni Formations, which consists of one or more of conglomerate, metabasalts, metapelites, slates, phyllites, arenite, ultramafics, cherts, banded iron formation (BIF), etc. See Table 4.

S 3 Salkhala Group (with Tanwal Formation): A Precambrian group of Higher Himalaya in Jammu and Kashmir State and in Himachal Pradesh. See Precambrian of Himalaya.

S 4 Salooni Formation: A fossiliferous Permian age formation of Himachal Pradesh, consisting of shale, slate and limestone.

S 5 Sancha Malla Group: See Malla Johar Supergroup and Table 17.

S 6 Sandhan Formation: A Piocene age formation of Kutch, Gujarat. See Table 38.

S 7 Sandmata Complex: It is the youngest formation of Bhilwara Supergroup of the Aravalli Craton and overlies The Mangalwar Complex. The boundary between the granulite belt of Sandmata Complex and the gneissic terrain of Mangalwar Complex is marked by Delwara Lineament. The contact appears to be gradational, marked by amphibolites-granulite facies transition. Sandmata Complex consists of large areas of migmatitic gneisses in amohibolite facies containing enclaves similar to those of Mangalwar Complex (called Tanwan Group) as seen at Amet. There are also enclaves that are similar to those in basement gneisses (Mewar Gneiss) called Sawadri Group. See Table 6.

S 8 Sangeluma Group: The Cretaceous rocks of Jammu and Kashmir are represented by Sangeluma Group and Hundri Formation, however these rocks extend much beyond Jammu and Kashmir, upto Tsangpo Valley in Tibet. The Sangeluma Group lies in the Tectonic Zone (Indus Suture) of Ladakh and includes Dras-Kargil and many other areas. The Sangeluma Group consists of sediments and lava flows and includes Khalsi Formation, Dras Formation and Dras volcanics. The Khalsi Formation consists of limestone, shale, phyllite, conglomerate, gritstone, chert beds, volcanogenic sediments, tuffaceous shale, minor quartz arenite and tectonic slices of ophiolite. The Dras Formation consists of black shale, volcano sedimentary, etc. and has been deformed, metamorphosed and intruded by diorites. Dras Volcanics consists of marine sediments, volcanic, pyroclastic hypabyssal and plutonic rocks. See Cretaceous of Jammu and Kashmir.

S 9 Sanglung Formation/Group: An Upper Triassic age formation of Spiti area in Himachal Pradesh. in Lilang Group the formation consists of Juvavites Beds, Tropites Beds and Grey Beds, while in Lilang Supergroup the Sanglung Group consists of Rongtong Formation with Lower, Middle and Upper Members. See Lilang Group and Lilang Supergroup and Tables 13 and 14.

S 10 Sanu Formation: A Paleocene to Eocene age fossiliferous formation of Jaisalmer Basin in Western Rajasthan, overlying the Habur Formation and under the Khuiala Formation. Sanu Formation is a sequence of sandstone and glauconitic clays passing upwards into coarse grained gritty sandstone. See Table 16.

S 11 Sargur Group: Overlies the Gorur Gneiss in the Western Dharwar Craton and consists of quartzite, garnet-biotite schist, para-gneisses, marbles, calc-silicate rocks, cherts, Banded Iron Formation (BIF), amphibolites and ultramafic-mafic intrusions. See Table 2.

S 12 Sarnu Formation: A Cretaceous age formation of Barmer Basin in Western Rajasthan, overlying the Jaisalmer Formation and under the Fatehgarh Formation. See Table 16.

S 13 Saser Brangsa/Shaksgam Formation: A Carboniferous age fossiliferous formation of Karakoram Range in Hunza and Ladakh area of Jammu and Kashmir and consists of argillite, limestone, sandstone, greywackes, etc.

S 14 Satnala Shale Formation: A formation of Penganga Group of Proterozoic age of Godavari Supergroup in the Pakhal/Mallampalli Belt. See Godavari Supergroup and Table 30.

S 15 Satpura mobile Belt: It is a long E-W trending Proterozoic age Mobile Belt which is divided from west to east into following 4 sectors: 1. Aravalli Craton with its Aravalli-Delhi orogen; 2. Central India tectonic Zone (CITZ); 3. Chhotanagpur Gneiss with North Singhbhum Orogen (NSO); and 4. Meghalaya Plateau of NE India. A tentative correction of the geological formations and structural features of the 4 sectors is given in Table 10. See the Table.

S 16 Satyavedu Formation: A Gondwana Supergroup formation of Middle Jurassic(?) age of Palar Basin in Tamil Nadu, consisting of conglomerate and sandstone.

S 17 Sauna Volcanics: A Precambrian formation of Lesser Himalaya in Jammu and Kashmir. Sauna Volcanics are associated with Ramban Formation. See Precambrian of Himalaya.

S 18 Saundatti Quartzite: See Kaladgi Basin and Table 37.

S 19 Sausar Supracrustal Belt/Group: A formation of Central Indian tectonic Zone (CITZ) of the Stpura Mobile Belt, consisting from bottom to top of Older Metamorphic Group, Tirodi Gneiss, Sitosaongi Formation, Lohangi Formation, Mansar Formation, Charbaoli Formation, Juniwani Formation and Bichua Formation.

S 20 Sawai Bay Formation: A Miocene to Pliocene age formation of Archipelago Group in the Andaman-Nicobar Islands. See Table 21.

S 21 Se La Group: A Precambrian group of Higher Himalaya in Arunachal Pradesh. See Precambrian of Himalaya.

S 22 Sedam Subgroup: See Bhima Basin/Group.

S 23 Semri Group: The oldest Group of the Vindhyan Supergroup. For details see Table 9.

S 24 Serpentine Group: See Cretaceous of Andaman and Nicobar Islands.

S 25 Seti Khola Formation: A Permian age formation of Bhutan, consisting of sandstones, shales and quartzwackes.

S 26 Shahabad Limestone : See Bhima Basin/Group.

S 27 Shal Shal Formation: A Triassic age formation of Niti area, Painkhanda, Uttarakhand consisting of fossiliferous, predominantly limestones and shales with quartzites in the Kioto Member. The Formation is divisible into Rambakot, Niti, Hoti, Rimkhim and Kioto Members. For details see Table 26.

S 28 Shali Group: A Precambrian Group of Lesser Himalaya in Himachal Pradesh: See Precambrian of Himalaya.

S 29 Shali Window: A structure in the second Precambrian sedimentary cycle of Lesser Himalaya in Himachal Pradesh, which exposes Sundergarh Group with Mandi Volcanics and Shali Group.

S 30 Shell Limestone: A Recent to sub-Recent age formation of Archipelago Group in the Andaman and Nicobar Islands consisting of coral and sand. See Table 21.

S 31 Shiala Formation: The middle formation, most probably of Ordovician age in the Ordovician-Silurian succession in the Kumau-Garhwal area of Uttarakhand.

S 32 Shillong Group: The Group overlies the Older Metamorphic Group which is an extension of Chhotanagpur Gneiss of the Satpura Mobile Belt in the NE India. The Group consists of 2 formations, lower the Mawphlang Quartzite and upper the Mawngup Phyllite. See Satpura Mobile Belt.

S 33 Shimoga Schist Belt or Super Belt: It is associated with Bababudan Group of the Dharwar Supergroup in the western Dharwar Craton, covering an area of 25,000 sq. km. and extends from Tarikere valley to Goa to beyond Ratnagiri in Maharashtra. See Tables 2 and 3.

S 34 Shumahal Formation: A formation of Hapatnar Group of Cambrian to Middle Silurian age in the Palaeozoic Succession of Lidder Valley in Kashmir. See Table 11.

S 35 Shumar Formation: A formation of Jaisalmer Basin in Western Rajasthan of Pleistocene to sub-Recent age. See Table 16.

S 36 Siju Formation: An Eocene age formation of Sylhet sub-Group of Jaintia Group, overlying the Tura Formation in the Garo Hills area of NE India. See Table 23.

S 37 Sillakudi Formation: A Formation of Ariyalur Group of Uppermost Cretaceous age in the Trichinopoly area of Tamil Nadu. See Table 18.

S 38 Silurian of Kashmir: Sandy shales near Anantnag contain Upper Silurian fossils. Silurian slates occur in Sindh valley overlying the Ordovician greywackes and slates. In Warwan valley of Doda district a thin band of fossiliferous Silurian rocks occur below the Ordovician Muth Quartzites and above the Cambrian Dogra Slaes.

S 39 Silurian of Kumaon, Uttarakhand: In Kumaon area Ordovician-Silurian rocks occur together and the formations from bottom to top are: Variegated Formation, Shiala Formation and Garbyang Formation, all of these are fossiliferous. The Ordovician-Silurian boundary lies within Shiala Formation.

S 40 Silurian of Spiti, Himachal Pradesh: Consists of Takche/ Manchap Formation which is of Late Ordovician to Middle Silurian age and consists of siltstone, shale, limestone and dolomite.

S 41 Simikeri Subgroup: See Kaladgi Basin.

S 42 Simla Slate/Group: A Precambrian group of Lesser Himalaya in Himachal Pradesh. See Precambrian of Himalaya.

S 43 Simlipal Basin/Group: A Group of Archean age, equivalent to the Dhanjori Group of Singhbhum Craton, consisting of basal quartz pebble conglomerate, orthoquartzite and felspathic quartzite. See Table 5.

S 44 Sincha Formation: A Precambrian formation of Lesser Himalaya in Jammu and Kashmir. See Precambrian of Himalaya.

S 45 Sindreth Group: A Group of Neoproterozoic age of Marwar Craton of the Aravalli Craton, consisting of conglomerate, quartzite, shale and phyllite with flows of amygdular basalts, resembling the Deccan Trap basalts. See Table 6.

S 46 Singhbhum Craton: Occupies about 50,000 sq.km. area in northeastern part of the Indian Peninsula and contain formations of Archean and Proterozoic age. See Table 5 for stratigraphy.

S 47 Sirmur Group: The Group includes Subathu, Dagshai and Kasauli Formations of Himachal Pradesh and Uttarakhand, ranging in age from Eocene to Early Miocene. See Table 22.

S 48 Sirohi Group: A Group Neoproterozoic age of Marwar Craton of Aravalli Craton, consisting of fine grained metamorphosed sediments of pelite-quartzite-marble suite consisting of quartz-mica schist, carbonaceous phyllite, dolomitic marbles and calc-silicatr rocks. See Table 6.

S 49 Sivaganga Beds: These are coastal beds of Lower Cretaceous age of Upper Gondwana Supergroup occurring in the Ramnathpuram area of Sivaganga district, Tamil Nadu. The beds are richly fossiliferous and consists of boulder beds and conglomerate followed by sandstones, grits and shales.

S 50 Sivaganga Formation: An Upper Gondwana Supergroup formation of Lr. Cretaceous age occurring in the Trichy area of Tamil Nadu. The formation consists of 3 members, namely the Kovandakaruchi Conglomerate, Kovandakaruchi Siltstone and Terani clay containing fossils.

S 51 Siwalik Group: The Group comprises of a. Lower, b. Middle and c. Upper Divisions and consists of fossiliferous sedimentary rocks of Miocene to

Pleistocene age, forming the Himalayan foot hills or the outer Himalaya, extending from the Indus River in the west to Irravady River in Myanmar in the east. The Siwaliks on the north are separated from the Lesser Himalaya by a major thrust, the Main Himalayan Thrust, while on the south the Siwaliks are separated from the Indo-Gangatic Plain by the Main Boundary Fault. The Siwalik outcrops are generally 10-12 km wide and form parallel ridges with deep vallies in between, called Duns. In India the Siwalik Group occurs in Jammu and Kashmir at Jammu Hills and Poonch-Mirpur, etc. areas; in Himachal Pradesh at Hoshiarpur-Jogindernagar, Dharamshala, Nalagarh, Haritalyanagar, etc.areas and in Uttarakhand at Kalagarh, Laldhang, Kotdwar, Ramnagar, etc. areas; in West Bengal at Jalpaiguri and Darjeeling districts and in Arunachal Pradesh all along the foot hills from border with Bhutan in the west to Roing in Dihang Valley in the east. See Table 28 for classification and lithology of the Siwalik Group.

S 52 Siwana Granite or Siwana Ring Complex: Is of Neoproterozoic age and belongs to the Malani Igneous Suite of the Marwar Craton of the Aravalli Craton. Siwana Granite consists of six granitic plutons covering an area of 1100 sq. km. The main ring complex has been formed by collapse of volcsanic caldera in about 800 sq.km. area, followed by intrusions of ring dykes of peralkaline granite along its periphery. The main magmatic phase of Siwana Granite with their basal felsic volcanics and mafic dykes are: (i) porphyritic granite, (ii) leucogranite and pegmatite and (iii) microgranite and granite porphyry. See Tables 6 and 16.

S 53 Sleemanabad Formation: The lowermost formation of Mahakoshal Group of the Satpura Mobile Belt. The lower part of the formation is also called agori Formation and consists of quartz arenite, stromatolitic carbonate rocks, BIF (Banded Iron Formation) and phyllite, while the upper part of the formation, also called Chitrangi Formation and consists of pillowed basalts, manganiferous chert, BIF and ultramafics. See Satpura Mobile Belt and Table 10.

S 54 Solda Formation: See Sonrai Basin

S 55 Somadevara Quartzite Formation: A formation of Venkatapuram Group of Albaka Belt of the Godavari Supergroup of Proterozoic age. See Godavari Supergroup and Table 30.

S 56 Somanpalli Group: Oldest formation of Albaka Belt of the Godavari Supergroup of Proterozoic age. See Godavari Supergroup and Table 30.

S 57 Sonakhan Group; A small outcrop of Proterozoic age, south of Chhattisgarh Basin in the Bastar Craton, consisting of Lower Baghmara Formation/Bilari Formation and upper Arjuni Formation with an unconaormity inbetween. See Table 4.

S 58 SONATA (Son-Narmada-Tapti) Lineament: Exists at the southern end of Aravalli Craton, forming the northern boundary of Central Indian Tectonic Zone (CITZ).

S 59 Sonrai Basin: A Paleoproterozoic age Purana Basin of Bundelkhand Craton in Madhya Pradesh.The basin houses the Sonrai Formation consisting of grits, shale and dolomite, followed by Kurat Basalt and Solda Formation. The Solda Formation consists of quartzites and shale. See Table 32.

S 60 Sonrai Formation: See Sonrai Basin.

S 61 Spiti Formation/Shale: The type area of Spiti Shale/Formation is the Lingti Valley in Himachal Pradesh, where the formation ranges in age from Upper Jurassic (Oxfordian) to Lower Cretaceous (Valanginian to Hauterivian) (see Table 13) and comprises of black splintery shale, fine grained sandstone and siltstone and occasionally gypseous scales and phosphatic nodules enclosing fossils. In earlier literature, the Spiti Shale Formation in the Spiti area was shown divisible into 3 beds, which from bottom to top are; Belemnopsis gerardi Bed, Chidamu Bed and Lochambal Bed. The Spiti Shale Formation is also found in the Malla Johar area near Soncha Malla in Pithoragarh area, Kumaon, Uttarakhand, where the formation is of Upper Jurassic to Lower Cretaceous age (see Table 17) and consists of shales, silts, nodules and ammonite fossils.

S 62 Spiti Member: A Member of Tithonian (Upper Jurassic) age of Kiogad Formation in the Niti area of Painkhanda, Uttarakhand, consisting of shales wirth calcareous nodules. See Table 26.

S 63 Sriperumbudur Formation: A Middle Jurassic (?) age formation of Palar Basin in Tamil Nadu, having affinity with the Gondwana Supergroup. See Table 12.

S 64 Strait Formation: A formation of Miocene age of Archipelago Group in stratigraphy of Andaman and Nicobar Islands. The formation is also called Round Formation. See Table 21.

S 65 Subathu Formation/Group: Oldest formation of Sirmur Group of Eocene to Miocene age occurring in Himachal Pradesh and Kumaon-Garhwal areas of Uttarakhand. Subathu formation is divisible into lower Member A, and upper Member B. Member A consists of 160 m of green shale, foraminiferal limestone and oyster marls, quartz arenite, carbonaceous shale and calcareous sandstone, while Member B consists of 12 to 90 m of red, green, indigo blue mudstone, shale and quartz arenite. The Eocene age rocks of Subathu Group of Kalakot-Metka-Moghala inliers of Rajauri district, J & K, contain vertebrate fossisls.

S 66 Suket Shales: It is the uppermost formation of Semri Group of Vindhyan Supergroup, occurring in the Chittorgarh area of Rajasthan and consisting of soft fossiliferous shales of different colours with ripple marks. See Table 9.

S 67 Sukma Group: oldest Group of Archean age in the Bastar Craton. The group consists from bottom to top of 1. Sillimanite quartzite, 2. Diopsidic calc-silicate rocks and para amphibolites, 3. Cordietite-andasulite-garnet-biotite schist, gneisses and quartzites, 4. Cordierite-anthopyllite rocks, 5. Amphibolites and ultramafics, 6. Banded iron formation (BIF) and 7. Ultrmafic and maafic rocks. See Table 4.

S 68 Sulcacutus Bed: A thin black ferruginous Oolite bed rich in Belemnites sulcacutus fossils of Upper Jurassic age in Spiti area of Himachal Pradesh, lying over the Tagling Limestone of the Kioto Limestone. Sulcacutus Bed probably forms part of Spiti Shale Formation and may not have any independent identity. Bed similar to Sulcacutus Bed is also found in Kumaon area of Uttarakhand, overlying the Laptal Formation.

S 69 Sullavai Group: Youngest Group of Proterozoic age Godavari Supergroup in the Pakhal/Mallapalli Belt. See Table 30.

S 70 Sundergarh Group: A Precambrian group with associated Mandi Volcanics of Lesser Himalaya in Himachal Preadesh. See Precambrian of Himalaya.

S 71 Supra Panchet Formation: An Upper Triassic Formation of Gondwana Supergroup, occurring in the Damuda and Son- Mahanadi Basins. The formation predominantly consists of ferruginous sandstone, conglomerate, pebble beds and red clay. Mahadeva and Maleri are its equivalent formations. See Table 12.

S 72 Surajdeval Formation: See Dharangadhara Group.

S 73 Surma Group: A Miocene age group of NE India in Assam, etc. areas, comprising of lower Bhuban and upper Bokabil Formations. See Table 22 and 23.

S 74 Suru Gneiss: A Precambrian gneiss of Higher Himalaya in Himachal Pradesh. See Precambrian of Himalaya.

S 75 Suwana Formation: An Archean age formation of Mangalwar Complex of Bhilwara Supergroup of the Aravalli Craton, overlying the Potla Formation and underlying the Kekri Formation. Suwana Formation which occurs as small Swathe within Potla Formation near Bhilwara, consists of migmatised amphibolites with minor bands of garnet-biotite-kyanite schist, calc-silicate rocks and quartzite. See Table 6.

S 76 Sylhet Limestone Formation: A formation of Jaintia Group of Paleocene-Eocene age overlying the Tura Formation, occurring in Mikir Hills-Dhansiri Valley and Upper Assam Valley of Assam-Arakan Basin in NE India. See Tables 22 and 23.

S 77 Sylhet Sub-group/Formation: A Sub-group of Jaintia Group of Eocene age in the Garo Hills, Khasia and Jaintia Hills areas of Assam-Arakan Basin in NE India. In Garo Hills the Sub-group lies over the Langarin Member and consistsof lower Tatu Formation and upper Siju Formation, while in Khasi and Jaintia Hills the Sub-group comprises from bottom to top of Lakadong limestone, Lakadong Sandstone, Umlatdoh Limestone, Narpuh Sandstone and Prang Members. See Table 23.

S 78 Sylhet Trap: Consists of volcanic rocks which are predominantly quartz theolities and are of Permian to Early Cretaceous age and occur in Garo, Khasi and Jaintia Hills of Assam-Arakan Basin in NE India. See Table 23.

S 79 Syringothyris Limestone: A Lower Carboniferous age formation of Lidder Valley near Aishmuqam in Kashmir, underlies the Passage Beds Formation (Aishmuqam Formation) and overlies The Tannaki Conglomerate Formation. Syringothyris Limestone has been named after the index fossil Syringothyris cuspidata which abundantly occurs in the Formation which consists of a series of limestone beds, mostly flaggy.

T 1 Tagling Limestone: The uppermost bed of Jurassic age of Lilang Group, Spiti, consisting of 500 m of fossiliferous massive limestone and dolomite. See Table 13.

T 2 Takalpalli Arkose Formation: Oldest formation of Penganga Group of the Godavari Supergroup of Proterozoic age in the Pakhal/Mallampalli Belt. See Table 30.

T 3 Takche/Manchap Formation: Belongs to Late Ordovician to Middle Silurian age and occurs in the Spiti Valley of Himachal Pradesh. The formation overlies the Thango Formation and underlies the Muth Quartzites Formation. Lithology of the formation varies from place to place, from sandstone to shale to limestone to dolomite.

T 4 Takli Formation: An Inter-trappean bed/formation of Nagpur area of Late Cretaceous age.

T 5 Tal Group/Formation: belongs to early Cambrian age and occurs in the Krol Belt of the Lesser Himalaya in Himachal Pradesh. Tal Group is divided into Shaliyan, Sankholi and Koti Dhaman Formations. In Mussoorie area of Uttarakhand, the Tal Formation is found overlying the Krol Formation. See Precambrian of Himalaya.

T 6 Talchir Formation: This is the oldest formation of Upper Carboniferous to Lower Permian age of the Gondwana Supergroup and overlies Precambrian basement. Talchir Formation consists of boulder bed/tillite, rhythmite, khaki green needle shale and light green sandstone. The formation is well represented in all the major coal fields of Damodar Valley, e.g. Daltonganj sub-basin, etc. areas. See Table 12.

T 7 Tamba Kurkus Group: A Triassic age Group of the Lilang Supergrgoup of Spiti in Himachal Pradesh. See Lilang Supergroup and Table 14.

T 8 Tanawal/Conglomerate/ Formation: A Precambrian formation of Higher Himalaya in Jammu and Kashmir and Himachal Pradesh, occurring with Salkhala Group. See Tannaki Conglomerate.

T 9 Tandi Group: This is a post-Cambrian Palaeozoic Group of Lahaul area of Himchal Pradesh- probably of Permian age. The Group is divisible into Kukti, Gushal and Dilburi Formations.

T 10 Tannaki/Tanawal/Langera Conglomerate: This is the oldest Carboniferous (may be Precambrian also) formation of Kashmir, consisting of boulder bed and conglomerate in the Bhadarwah-Bhallesh area.

T 11 Tarakeshwar Formation: An Oligocene age sub-surface formation of Gujarat Mainland. See Table 20.

T 12 Tatrot Formation: A Siwalik Group formation of Pleistocene age. See Siwalik Group and Table 28.

T 13 Tectonic framework of India; The Precambrian Indian Shield consists of cratons and mobile belts. Cratons have remained tectonically stable for prolonged periods and were affected only by younger epeirogenic movements and are dominated by Precambrian crystalline rocks, overlain by younger rocks. Mobile Belts are curvilinear, high-grade-gneissic-granulite belts surrounding the cratons, but separated from cratons by crustal scale ductile shear zones. The Indian Shield consists of 5 cratons of Archean age and 3 Mobile Belts of Proterozoic age. The 5 cratons are: Dharwar, Bastar, Singhbhum, Bundelkhand and Aravalli Cratons. The 3 Mobile Belts are: 1. Eastern Ghats Mobile Belt fringing Dharwar, Bastar and Singhbhum Cratons; 2. Pandyan Mobile Belt fringing Dharwar Craton; and 3. Satpura Mobile Belt fringing Bastar, Singhbhum, Bundelkhand and Aravalli Cratons and extending into Shillong Plateau. In addition to cratons and mobile belts, the Indian Shield also have a cover of younger rocks, which include: 1. The Proterozoic sedimentary basins, called Purana Basins in Indian Geology; 2. Deccan Traps or the Deccan Volcanic Province in western and central India with its outliers at Rajahmundry in southern India; 3. Rajmahal Taps in eastern India; 4. Coal bearing Gondwana successions in Damodar, Satpura, Mahanadi and Godavari rifts; 5. Cretaceous marine incursions in Narmada Valley and at Tiruchirapalli (Trichinopoly) in Cauvery Delta; 6. Coastal Gondwana Tertiary and Quaternary sediments with local volcanics; 7. Holocene (Recent) sediments along major river-vallies; Tertiary and Quaternary sediments at the Himalayan foothills and in Indo-Gangatic and Brahmaputra plains, including Bengal fan and 8. Phanerozoic succession in Himalaya.

T 14 Terani plant beds: Lower Cretaceous, Upper Gondwan beds of Trichinopoly area, with plant fossils. See Trichinopoly area and Table 18.

T 15 Teri sands: See Quaternary deposits of Tamil Nadu.

T 16 Teris: These are near coast Quaternary deposits of Kerala, consisting of peat beds, sand bars, sandy flat, raised sand beaches and reddish sandy loam containing heavy minerals like ilmenite, zizrcon and hematite.

T 17 Tertiary: The term Tertiary does not find mention in the International Stratigraphic Chart 2004 issued by International Commission on Stratigraphy. However the term is still in use in the Indian Stratigraphy to indicate time period from Paleocene to pliocene. See Cenozoic Era and Table 19.

T 18 Tertiary of Gujarat: See Table 20.

T 19 Tertiary of Kashmir: Tertiary formations in Jammu and Kashmir are best represented by Eocene rocks, which predominantly consists of quartzite and shale. However Miocene rocks also occur, which are represented by Murree Formation.

T 20 Tethyan Basin: A stretch of Himalayan terrain, north of Main Central Thrust extending from Kashmir to Arunachal Pradesh and occupied by Palaeozoic and younger rocks. The Tethyan Basin includes 5 sub-basins, namely: 1. Kashmir, 2. Bhadarwa-Chamba, 3. Zanskar-Spiti-Kinnaur, 4. Garhwal-Kumaon, and 5. Nepal-Sikkim-Bhutan-Arunachal Pradesh.

T 21 Thabo Beds: Lower Carboniferous beds of Po Formation of Kanwar Group of Spiti area in Himachal Pradesh, consisting of black shales with fossils of ferns and allied plant leaves. Due to its small area and thickness it has probably lost its independent identity.

T 22 Than Formation: See Dharangadhara Group.

T 23 Thango Formation: An Ordovician formation of Himachal Pradesh, consisting of conglomerate, quartzite, siltstone and shale and is well exposed in the Parahio river section in Spiti. The formation overlies the Kunzum La Formation and underlies the Takche Formation.

T 24 Therria Formation: A Paleocene age formation of Khasi and Jaintia Hills, Shillong Plateau in Assam and Meghalaya in NE India, consisting of 3 Members, which from bottom to top are: Therria Limestone, Cherra and Therria Sandstone. See Tables 22 and 23.

T 25 Thimpu Group: A Precambrian Group of Lesser Himalaya in Sikkim-Darjeeling. See Precambrian of Himalaya.

T 26 Tikak Parbat Formation: Uppermost formation of Barail Group of Oligocene age in the Nagaq Hills Fold Belt of Assam-Arakan Basin of NE India. See Table 23.

T 27 Tikki Formation: A Lower Triassic age formation of Gondwana Supergroup in the Rewa area of Son-Mahandi Basin, consisting of white fine grained sandstone with some siltstone. See Table 12.

T 28 Tindra Formation: A Lower Triassic age formation of Central Kashmir Valley, Pastannah area, consisting predominantly of nodular limestone, limestone with few nodules and subordinate limestone.

T 29 Tipam Formation/Group: A Pleistocene-Miocene age formation of Khasi and Jaintia Hills, Mikir Hills and Dhansari Valley, Surma Valley and Tripura Fold Belt areas of Assam-Arakan Basin in NE India, consisting of massive sandstones with thin bands of shale and clay. See Table 23.

T 30 Tipam Sandstone Formation: The lowermost part of Tipam Formation in the upper Assam Valley. See Table 23.

T 31 Tippapuram Shale Formation: A formation of Venkatapuram Group of Albaka Belt of the Godavari Supergroup of Proterozoic age. See Table 30.

T 32 Tiratgarh Formation: Lowermost formation of Indravati Basin, see Table 36.

T 33 Tirodi Gneiss: A biotite gneiss of granodioritic to granitic composition, occurring north of Sakoli Belt and is in tectonic contact with Amgaon Gneiss of Bastar Craton.

T 34 Tirupati Formation: A Gondwana Supergroup formation of Lower Cretaceous age, occurring in Krishana-Godavari coastal area of Andhra Pradesh, overlying Raghavapuram Formation and consisting predominantly of conglomeratic/ferruginous/micaceous sandstone and claystone with Ptilophyllum flora and fossiliferous limestone (Infra-trappean). See Table 12.

T 35 Trace Fossils: See Ichnology.

T 36 Traumatocrinus Beds: Triassic age beds of Painkhanda, Uttarakhand. The term has probably lost its identity and stands merged with the Rimkhim Member of the Shal Shal Formation.

T 37 Triassic of Byans: In Byans, Uttarakhand, Triassic rocks occur in Kumaon where these have been divided into Chocolate Series, Kalapani Limestone, Kuti Shales and Kioto Formation.

T 38 Triassic of Jammu and Kashmir: In Kashmir Valley, Triassic formations from bottom to top are Khunamuh, Tindra, Khreuh, Wuyan and Wumuh Formations. See each. In Jammu Triassic rocks occur in association with Zewan Beds in the outer Panjal Trap belt in poonch-Banihal area. In Ladakh Upper Triassic rocks occur in the Tibetan side of the Himalaya range and consists of limestone and dolomite.

T 39 Triassic of north Kumaon and Garhwal: See Triassic of Painkhanda and Byans.

T 40 Triassic of NW Himalaya: Marine life Triassic age fossils were recorded in certain areas of NW Himalaya, which include Spiti, Kiogarh Valley, Kashmir, Ladakh, Niti Pass, Painkhanda (Kumaon, Uttarakhand).

T 41 Triassic of Painkhanda: See Table 26.

T 42 Triassic of Spiti: See Tables 13 and 14.

T 43 Trichinopoly (Tiruchirapalli) area: Lies in the eastern part of Tamil Nadu and includes Pondicherry, Virdhachalam and Ariyalur areas. See Table 18 for stratigraphy.

T 44 Trichinopoly Group: See Trichinopoly area and Table 18.

T 45 Trigonia Beds: Cretaceous age beds of Umia Formation, Kutch, Gujarat. See Table 15.

T 46 Trigonoarca Beds: Cretaceous age beds of Vadavur, Pondicherry. See Cretaceous of Pondicherry.

T 47 Tropites Beds: Upper Triassic age beds of Lilang Group, Spiti, Himachal Pradesh, consisting of 100-200 mts of fossiliferous limestone and shales. Beds equivalent to Tropites Beds also occur in Byans, Uttarakhand. See Table 13.

T 48 Tso Morari Gneiss: A Precambrian gneiss of Higher Himalaya in Himachal Pradesh. See Precambrian of Himalaya.

T 49 Tura Formation: A formation of Jaintia Group of Paleocene-Eocene age, occurring in the Garo Hills, Mikir Hills and Dhansari Valley and Upper Assam Valley of Assam-Arakan Basin in NE India. See Table 23.

T 50 Tusham/Tosham Ring Complex, Haryana: Belongs to the Malani Igneous Suite of the Aravalli Craton and consists of rhyolite and granite intruding andalusite bearing carbonaceous mica schist of Sirohi Group. The ring complex is known for its unique Cu-(Au)-Ag-Sn-Bi-In mineralization. In is for Indium.

U 1 Udaipur Group: A Group of Aravalli Supergroup, overlying the Debari Group and consisting of alternate beds of gradded bedded greywacke and laminated phyllite with intercalations of ferruginous dolomitic chert, phosphatic algal dolomite and carbonaceous to magniferous phyllite. The greywacke phyllite suite is overlain by Mochia/Zawar Formation which for many years at Zawar Mines, near Udaipur was the only source for Pb-Zn-Ag ores in the country.

The Mochia/Zawar Formation consists of dolomites associated with carbonaceous phyllites and quartzite. The Mochia/Zawar Formation is overlain by Bowa Formation consisting of phyllites and quartzose phyllites in which sedimentary structures are well preserved. Machala Magra and Tidi Formations consisting of phyllite quartzite sequence were formed due to dolomitisation in the Zawar Formation. See Table 6.

U 2 Ukra Beds: Cretaceous age beds of Umia Formation, Kutch, Gujarat. See Table 15.

U 3 Ukra Member: See Umia Stage.

U 4 Um Sohryngkew Formation: A 250 m thick formation of Late Cretaceous age of Meghalaya Group in NE India. See Meghalaya Group.

U 5 Umaria Marine Bed: About 3m thick marine bed of Permian age, consisting of sandstone and clay, occurs intercalated in the Lower Gondwana strata in the Narsarha railway cuttings near Umaria in Madhya Pradesh. Similar intercalations occur at Anukpur and Mahendragarh about 150 km ESE of Umaria. The umaria Marine Bed overlies the Talchir Boulder Bed of the Gondwana Supergroup and passes gradually into the overlying Barakar rocks. Fossils of Productus, Spiriferina, Reticularia and Athyris are found in the Umaria Marine Bed.

U 6 Umia ammonite Bed: Lowermost bed of Umia Formation of Cretaceous age of Kutch, Gujarat. See Table 15.

U 7 Umia Beds: Belongs to the Umia Formation of Cretaceous age, of Kutch, Gujarat. See Table 15.

U 8 Umia Formation: A Cretaceous age formation of Kutch, Gujarat, consisting of many beds. Upper part of Umia Formation (Bhuj Beds) contain plant fossils comparable to some of the Upper Gondwana formations. See Table 15.

U 9 Umia Member: See Umia Stage.

U 10 Umia Plant Beds: Upper beds of Bhuj Beds, Kutch, of Cretaceous age containing a few plant fossils. See Table 15.

U 11 Umia Stage: In the Mesozoic succession of Kutch, excluding the Bhuj Beds and Ukra Beds, rest of the Umia Formation makes the Umia Stage, extending in age from Tithonian to Neocomian of Crtaceous Period. The beds of the Umia Stage from bottom to top are Umia ammonite bed, Barren sandstones, Trigonia beds, Barren sandstones and shales and Umia beds. Biswas in 1971 suggested four fold classification of Umia Stage. The four Members from bottom to top are Umia or Katesar, Ghuneri, Ukra and Bhuj. See Table 15.

U 12 Umlatdoh Limestone Member: A Member of Sylhet Subgroup of Jaintia Group of Eocene age, occurring in the Khasi and Jaintia Hills of Assam-Arakan Basin in NE India. See Table 23.

U 13 Upper Assam Basin: One of the Basins producing hydrocarbons in India. See Table 29.

U 14 Upper Katrol Sandstone: A bed of Katrol Formation of Jurassic age in Kutch, Gujarat. See Table 15.

U 15 Upper Katrol Shales: Uppermost bed of Katrol Formation of Jurassic age in Kutch, Gujarat. See Table 15.

U 16 Upper Muschelkalk: A Middle Triassic age bed of Lilang Group, Spiti, Himachal Pradesh. See Table 13.

U 17 Upper Sandstone Formation: A Pliocene age formation of Tipam Group, occurring in the Upper Assam Valley of Assam-Arakan Basin in NE India.

U 18 Usur Group: Uppermost Group of Albaka Belt of Godavari Supergroup of Proterozoic age. See Table 30.

U 19 Uttapalli Conglomerate Formation: A formation of Penganga Group in the Albaka Belt of Godavari Supergroup of Proterozoic age.

U 20 Uttarlai Formation: A Quaternary age formation of Barmer Basin in western Rajasthan, consisting of dune sands and alluvium gravel/trap wash. See Table 16.

U 21 Uttatur Group: A Cretaceous age Group of Trichinopoly area. See Table 18.

V 1 Vagadkol Formation: A Paleocene age unaossiliferous formation of Gujarat Mainland, overlying Deccan Traps and consisting of conglomerate and variegated clays derived from weathering of basalt. See Table 20.

V 2 Vaikom Formation: Considered to be a part of Quilon Formation of Kerala of Miocene age and consists of gravel, sand, clays and thin lignite seams. See Table 25.

V 3 Vaikrita Group: A Precambrian Group of Central Crystallines of Higher Himalaya in Uttarakhand. See Precambrian of Himalaya.

V 4 Valudavur Beds: See Cretaceous of Pondicherry.

V 5 Variegated Formation: The lowermost formation of the Ordovician-Silurian succession in the Kumaon-Garhwal area of Uttarakhand, underlying the Shiala Formation:

V 6 Vejendla-Vemavaram Formation: The Cretaceous age Upper Gondwana Formation of Vejendla-Vemavaram area in Guntur and Prakasam districts of Andhra Pradesh, consisting of sandstones grading into shale. Earlier named formations, the Badavada Sandstone, Vemavaram Shale and Pavalur Sandstone are included into the Vejendla-Vemavaram Formation.

V 7 Vemavaram Formation: See Vejendla-Vemavaram Formation:

V 8 Vembanad Formation: The uppermost formation of Quaternary age of the Malabar Supergroup of Kerala, consisting of Beach sands, alluvium, red (teri) sands, peat beds, clays, etc. See Table 25.

V 9 Venkatapur Sandstone Formation: The uppermost formation of Sullavai Group of Pakhal/Mallampalli Belt of Godavari Supergroup of Proterozoic age. See Table 30.

V 10 Venkatapuram Group: A Group of Albaka Belt of Godavari Supergroup of Proterozoic age. See Table 30.

V 11 Vindhyan Basin: It is one of the largest Proterozoic Sedimentary Basins (Purana Basins) of Meso-Neoproterozoic age in the Bundelkhand Craton and houses the Vindhyan Supergroup. See Vindhyan Supergroup and Table 32.

V 12 Vindhyan Supergroup: The Vindhyan Supergroup of Meso-Neoproterozoic age occupies the Proterozoic Purana Vindhyan Basin in the Bundelkhand Craton. The Supergroup consists of 4 groups, which from bottom to top are- Semri Group, Kaimur Group, Rewa Group and Bhander Group. See each Group and Table 9.

V 13 Vinjhan Shale Formation: Upper formation of Miocene ag of Vinjhan Stage in Kutch, Gujarat.

V 14 Vinjhan Stage: A Miocene age Stage of Kutch, Gujarat, also encountered sub-surface in Lakhpat-I Well, Kutch, Gujarat, consisting of fossiliferous gypseous clay with hard marl bands. The Stage is equivalent to Khari Series of Kutch. See Table 20.

W 1 Wadhwan Formation: See Dharangadhara Group.

W 2 Waior Stage: An Oligocene age stage of Bermoti Series of Kutch, Gujarat, overlying the Ramania Stage. The Waior Stage is also encountered in Lakhpat-I Well, Kutch, Gujarat. See Table 20.

W 3 Wangtu Gneiss: A Precambrian gneiss of Lesser Himalaya in Himachal Pradesh. See Precambrian of Himalaya.

W 4 Warkalli Group: The Group comprises the Tertiary formations of Kerala, which from bottom to top are Mayyanad, Quilon and Ambalapuzha Formations, consisting of sandstone, marls, limestone, clays. lignite, etc.

W 5 Wazur Formation: A Devonian age formation of Chorgali Group of Lidder Valley, Kashmir. See Table 11.

W 6 Wumuh Formation: An Upper Triassic to Lower Jurassic age formation of Central Kahmir Valley, Pastannah area, consisting of a few thousand metre thick massive limestone, divisible into Megalodon Limestone Bed and Halobia Limestone Bed.

W 7 Wuyan Formation: An Upper Triassic (Carnian-Norian) age limestone formation of central part of Kashmir valley in Pastannah area, Tral valley.

Y 1 Yadahalli Argillite:A formation of Lokapur subgroup of Kaladagi Basin of Proterozoic age. See Kaladgi Basin and Table 37.

Y 2 Yamne Formation: An Upper Permian age formation of Dalbuing area of Arunachal Pradesh, consisting of about 5m of pale brown ferruginous shale and containing marine invertebrate fauna of Upper Permian age.

Y 3 Yargatti Argillite: See Kaladgi Basin and Table 37.

Y 4 Yashwantanagar Formation: The lower formation of Eastern Dharwar Craton of Dharwar Supergroup of Archean age, consisting of mafic-ultramafic rocks/ quartzites and Banded Iron Formation. See Tables 2 and 3.

Y 5 Yerrapalli Formation: See Pranhita-Godavari Valley.

Y 6 Yinkiang Group: An Upper Paleocne to lower Eocene age Group of Arunachal Pradesh, consisting of Geku Formation and Dalbuing Formations, unconaormably overlying the Abor Volcanic Formation of Lower Permian age with following details. See Table 22 also.

Age	Formation	Lithology
Eocene	Dalbuing Formation	Quartzite, sandstone, shale (fossiliferous)
Paleocene	Geku Formation	Welded tuffs (? Ignimbrite), siltstone, shale, vesicular amygdaloidal mafic volcanics and shale with plant fossils.

Z 1 Zamia Beds: Lowermost beds of Bhuj Beds/Membe in the Cretaceous succession of Kutch, Gujarat. See Table 15.

Z 2 Zamia Shales: Same as the Upper Katrol Shales, the uppermost member of Katrol Formation of Jurassic age in Kutch, Gujarat. See Table 15.

Z 3 Zawar/Mochia Formation: See Udaipur Group.

Z 4 Zewan Formation: An Upper Permian age formation of Lidder Valley in Kashmir, Ladakh and Eastern Korakoram in J & K, consisting of varities of limestone, shale and sandstone. See Table 11.

■■■■

IV

LIST OF TABLES

Table No.	List of Tables
1	Geological Time Scale
2	Dharwar Craton
3	Dharwar Supergroup
4	Bastar Craton
5	Singhbhum Craton
6	Precambrian of Aravalli Craton
7	Cuddapah Basin
8	Delhi Supergroup
9	Vindhyan Supergroup
10	Satpura Mobile Belt
11	Paleozoic of Lidder Valley, Kashmir
12	Gondwana Supergroup
13	Stratigraphy of Spiti
14	Lilang Supergroup, Spiti
15	Mesozoic of Kutch, Gujarat
16	Stratigraphy of Western Rajasthan
17	Malla Johar Supergroup, Kumaon, Uttarakhand
18	Stratigraphy of Trichinopoly Area
19	Cenozoic Era
20	Cenozoic Stratigraphy of Gujarat
21	Stratigraphy of Andaman-Nicobar Islands
22	Stratigraphy of sub-Himalayan Region
23	Stratigraphy of NE India
24	Karewa Group, Kashmir Valley
25	Stratigraphy of Kerala
26	Stratigraphy of Painkhanda (Niti Area), Uttarakhand
27	Quaternary Sediments of Kali Valley, Kumaon, Uttarakhand

28	Siwalik Group
29	Basins Producing Hydrocarbons
30	Godavari Supergroup
31	Important Geological Units (Formations, etc.) of India
32	Purana (Proterozoic) Sedimentary Basins
33	Chhattisgarh Group
34	Strtigraphy of Bijawar Group, M.P. Bundelkhand Craton
35	Stratigraphy of Pari Group, Khari Basin, M.P. Bastar Craton
36	Stratigraphy of Indravati Basin, M.P., Bastar Craton
37	Stratigraphy of Kaladgi Basin (Proterozoic age) Dharwar Craton
38	Tertiary Stratigraphy of Kutch, Gujarat

Table 1. Geological Time Scale (Abbreviated)

Eon	Era	Sub-Era	Period	Epoch	Aga Ma
Pnaerozoic	Cenozoic	Quaternary	Neogene	Holocene (Recent)	0.0115
				Pleistocene	1.806
		Tertiary		Pliocene	5.332
				Miocene	23.03
			Paleogene	Oligocene	33.9
				Eocene	55.8
				Paleocene	65.5
	Mesozoic		Cretaceous		145.5
			Jurassic		199.6
			Triassic		261.1
	Paleozoic		Permian		299
			Carboniferous		359.2
			Devonian		416
			Silurian		443.7
			Ordovician		488.3
			Cambrain		542
Pre-cambrian	Proterozoic				2500
	Archean				(4000 or more, probably 4500 Ma)

Time unit = Equivalent rock unit Eon - Enothem Era = Erathem Period = System Epoch = Series Age = Stage Note : Ma = Million years ago

Table 2. Dharwar Craton

Age	Western Dharwar Craton (WDC)		Eastern Dharwar Craton (EDC)
Paleo-protero	Younger granite (Chitradurga, Arsekere granites) Charnockite		Younger granite/ gneiss (Closepet Granite and equivalents) Charnockite
Archean	Chitradurga Group Bababudam Group	DHARWAR SUPERGROUP	Kolar Group Yashwantnagar Formation
	---------------------- Unconformity----------------------		
	Peninsular Gneiss Sargur Group Gorur Gneiss		Enclaves of older gneiss ? Warangal Group ? Salem Group Putative Basement

Table 3. Dharwar Supergroup (Archean)

Western Dharwar Craton (WDC)		Eastern Dharwar Craton (EDC)	
Chitradurga Group	Ranibennur Subgroup Vanivilas Subgroup	Kolar Group	Champion Gneiss Goldfield Volcanics Sakarsanthalli Formation
------------------------------------ Disconformity ------------------------------------			
Bababudan Group	Mulaingiri Formation Santaveri Formation Allampura Formation Kalasapura Formation	Yashwantnagar Formation (Bababudan Group)	Mafic-ultramafic rocks/ quartzites Banded Iron Formation
Deformed angular unconformity		Basement not seen	

Table 4. Bastar Craton

Kotri-Dongargarh Orogon	AGE	Ma	Rest of the Crato
Chilpi Group and eqivalent Sakoli Group	Proterozoic	1800-2000	Chilpi Group = Sakoli Group
Khairagarh Group			
Dongarh Granite		- 2300	Malanjkhand Granite
Abujhmar Group			Kanker Granitoids
Nandgaon Group		2300	Sonakhan Group
Bailadila Group		2400	
Bengpal Gropu	Archean	2500-2600	
		2600	Bhopalpatnam and Kondagaon granulite Belts
			Sukma Granites
Amgaon Group		-3000	Sukma Group
> 3000 Gneissic complex with vestiges of 3500-3600 Ma events			

Table 5. Singhbhum Craton

Age	Present Classification	Near equivalents of previous work
Proterozoic	Newar Dolerite of different ages Kohlan Group (Kunjar, Gangpur Group) Kamakhyanagar Quartzite Soda granite, granophyre, Kuilapal granite Jagannathpur, Malangtoli Volcanics Dalma volcanic Group Mayurbhanj Granite Singhbhum Group / Granite and Pala Lahara Gneiss	Newar Dolerite Kohlan Series
Archean	Kojra Group with Ongarbira volcanics Dhanjori Group (=Simplipal Group)	Soda granite, granophyre and Arkasani Granite Singhbhum granite and Pala Laara Gneiss Dalma and Dhanjori lavas
	---------------------- Unconformity ----------------------	
	SBG-B, Bonai Granite, Chakradharpur Granite Tamper Kola granite, Mayurbhanj Granite Badampahar Group (=Gorumahasani Group) OMTG/SBG-A/Kaptipada granite and parts of Bonai and Chakradharpur granite OMG	Iron Ore Stage/Group Chaibasa Stage Singhbhum Granite basement
	Base note exposed	Basement not exposed

Notes: OMTG = Oder Metamorphic Tonalite Gneiss, OMG=Older Metamorphic Group, SBG = Singhbhum Granite

Table 6. Precambrian of Aravali Craton

Age			
Neoproterozoic	Marwar Craton 1000-542 Ma		Marwar Supergroup Malani Igneous uite Sirohi, Sindrath and Punagarh Groups
Mesoproterozoic		Delhi Supergroup 1600-1000 Ma	North Delhi Fold Belt Kishengarh Nepheline Syenite South Delhi Fold Belt
Paleoproterozoic		Aravalli Supergroup 2200-1800 Ma	Igneous intrusion Salumber (Dudas) granite Champaner Group Lunavada Group and Rikhabdeo ultramafics

			Jharol Group Udaipur Group Debari Group Delwara Group
Archean		Mineralised Supracrustal Belts	Bhinder Belt Rampur Agucha Belt Rajpur Dariba Belt Pur-Banera Belt Sawar Belt Jahazpur Belt
		Bhilwara Supergroup	Sandmata Complex Mangalwar Complex
		2900-2600 Ma	: Kekri Formation with Ranthambor Group Rajmahal Quartzite/Formation Suwana Formation Potla Formation Lasaria Formation Hindoli Group, including Berach Granite
		Mewar Gneiss - with 3300-3000 Ma	Jagat Group (assemblage of supracrustal rock fragments)

Table 7. Cuddaph Basin

Kurnool Group 200-450 m	Kundair Subgroup	Nandyal Shale Koilkuntla Limestone Paniam Quartzite
	Jammalamaduga Subgroup	Auk Shale Narji Limestone Banganapalli Quartzite
Cuddapah Supergroup	Nallamalai Group 3500-600 m	Srisailam Quartzite Cumbum Formation Pullampet Formation Bairen Konda Quartzite Nagari Quartzite
	Chitravati Group 4900-5000 Ma	Gandikota Quartzite Tadpatri Formation Pulivendla Quartzite
	Papaghni Group 2100 Ma	Vempalle Formation with Kuppalapalle volcanics Gulcheru Quartzite
---------------------------- Crystalline Basement of Dharwar Craton ----------------------------		

Table 8. Delhi Supergroup

<table>
<tr><td rowspan="7">Delhi Super group</td><td rowspan="3">North Delhi Fold Belt</td><td>Khetri Basin</td><td>North Khetri Basin
South Khetri Basin</td><td>Granite
Khetri Group
Granite
Shyamgarh Group</td></tr>
<tr><td>Alwar Basin</td><td></td><td>Granite
Ajabgarh Group
Alwar Group
Raialo Group</td></tr>
<tr><td>Bayana-Lalsot Basin</td><td></td><td>Ajabgarh Group
Alwar Group
Raialo Group</td></tr>
<tr><td colspan="4">Kishangarh Nepheline Syenite</td></tr>
<tr><td rowspan="3">South Delhi Fold Belt (SDFB)</td><td colspan="2">Albitite
Granite intrusions of the SDFB
Associate Group of SDFB</td><td></td></tr>
<tr><td colspan="2">Sendra Basin (West)</td><td>Sendra Group
Barotiya Group
Basantgarh Group</td></tr>
<tr><td colspan="2">Bhim Basin (East)</td><td>Bhim Group
Rajgarh Group</td></tr>
</table>

Table 9. Vindhyan Supergroup

Group (Thickness in meters)	Formation	Other names
Bhander Group 1300-1500	Maihar Sst Sibhu Sh Bundi Hill Sst Lakheri Lst Ganugarh Sh	Up Bhander Sst (Bavpura Sh Balwan Lst & Shikoada Sst) Lr Bhander Sst Bhander Lst, Nagod Lst Simraval Sh
-------------------------- Disconformity/Gradational contact --------------------------		
Rewa Group 100-300	Govindgarh Sst Drummondganj Sst Jhiri Sh Asan Sst Panna Sh	Up Rewa Sst Variegated Sh Lr Rewa Sst, Itwa Sst, Kanar Sst

-------------------------- Normal contact / Facies change --------------------------		
Kaimur Group 400	Dhandraul Quartzite	Up Kaimur Sst, Scrap Sst
	Mangesar Formation	
	Bijaigarh Sh	
	Markundi Sst	Ghaghar Sst
	Gurma Sh	Susnai Breccia
	Sasaram Sst	Lr Kaimur Sst
-------------------------- Unconformity / Normal contact --------------------------		
Semri-Group 3000-4000	Suket Sh	Baghwar Sh
	Rohtas Lst	Nimbahera Lst (Bari Sh, Jiram Sst, Khori Malan con)
	Chorhat Sst	Glauconite Bed, Rampur Sst, Basuhar Sst, Tirohan Breccia
	Bargawan Lst	Salkhan Lst, Fawn Lst, Chorhat Lst, Tirohan Lst
	Kheinjua Sh	Olvie Sh, Koldaha Sh, Binota Sh
	Kajrahat Lst	Kuteshwar Lst, Bhagwanpur Lst, Tirohan Lst, Lohar Dol
	Arangi Sh	--
	Deoland Sst	Khardeola Sst, Pandwafall Sst
-------------------------- Angular unconformity --------------------------		
-------------------------- pre-Vindhyan granites and supracrustals --------------------------		

Sst = Sandstone, Sh = Shale, Lst = Limestone, Dol = Dolomite, Por = Porcellanite, Con=Conglomerate, Lr=Lower, Up = Upper

Table 10. Satpura Mobile Belt

Aravalli Craton	C I T Z	Chotanagpur Gneiss/ NSO	NE India
Vindhyan basin	Vindhyan/Bijawar Basins	Munger Group	Shillong Group
GBF	SNNF	SNNF	Brahmaputra Fault
Hindoli			
Jahazpur Thrust	Mahakoshal Group SNNF	Mahakoshal Group SNSF	
Mangalwar Complex	Makrohar Granulite Belt	Bihar Mica belt	Gneiss with OMG
Bhilwara Group	Balrampur Fault		
Rajpur-Dariba Belt etc.	Betul Group	Hasathu-Belbatham belt	x
Kaliguman Shear Zone	Tan shear zone	North Purulia Shear Zone	
Sandamata Complex	Ramakona-Kalangi granulite Belt	Hazaribagh-Dumka granulite belt	x
Lunavada - Champaner	Sausar belt	Gangpur, Kunjar, Ranchi Groups	x
Delhi Supergroup			
Pilwa-Chainwali	Balaghat-Bhandara	Gumla-Purulia granulite belt	x

granulite basalt	granulite belt		
Phulad shear zone	CISZ	Tamar-Poropahar Shear zone	Phanerozoic Dauki Fault
Aravalli belt of Mewar Craton	Sakoli (?) and Chilpi Groups on Bastar Craton	Singhbhum Group of Singhbhum Craton	x

GBF = Great Boundary Fault; CITZ = Central Indian Tectonic Zone; SNNF = Son-Narmada North Fault; Central Indian Shear Zone; OMG = Older Metamorphic Group

Table 11. Paleozoic of Lidder Valley, Kashmir

Age		Group	Foramtion (Fm) (Thickness in Meters)
Mesozoic (Triassic)		Sonmarg Group	
Paleozoic Era	Upper Permian		Zewan Fm
	Lower Permian		Mamal Fm Panjal volcanic Fm Nishatbagh Fm
	Lower Permian to Carboniferous	Lidder Group	Agglomeratic Slate Fm/Pindobal Fm (700) Fenestella Shale / Ganeshpur Fm (1800) Pasage Beds / Aishmuqam Fm (700) Syringothyris Limestone Fm
	Devonian	Chorgali Group	Wazur Fm (500) Muth Quartzite (1000)
	Cambrain to Middle Silurian	Hapatnar Group	Rishkobal Fm (500) Rangamal Fm (400) Shumahal Fm (3500) Batal Fm and Kunzunla Fm in Zanskar area
--------------------------- Base not exposed ---------------------------			

Table 12. Gondwana Supergroup

Division	Age	Formation (Fm)
Upper Gondwana	Lower Cretaceous	Rajmahal Traps/Sediments, Bansa-Chandiya Fm, Jabalpur Fm, Chikiala Fm, Gangapur Fm Athgarh Fm, Golapalli Fm, Raghvapuram Fm, Tirupati Beds, Umia Fm
	Jurassic	Kota Fm, Marine Jurassic Formations of Kutch, i.e. Katrol, Chari and Patcham Formation Sriperumbudur Fm
	Upper and Middle Triassic	Dubrajpur Fm, Supra-Panchet Fm, Mahadeva Fm, Parsora Fm, Bagra Fm, Denwa Fm, Malaeri Fm

Lower Gondwana	Lower Triassic	Panchet Fm, Kamthi Fm, Tiki Fm, Pachmarhi Fm, Upper Member of Pali Fm
	Upper Permian	Raniganj Fm, Lower Kamthi Fm, Tiki Fm, Pachmarhi Fm, Upper Member of Pali Fm
	Middle Permian	Barren Measures, Lower Member of Pali F, Motur Fm
	Lower Permian	Barakar Fm, Karharbari Fm
	Upper Carboniferous to Lower Permian	Talchir Fm

Table 13. Stratigraphy of Spiti

<table>
<tr><td colspan="2">Age</td><td colspan="4">Stratigraphic units</td></tr>
<tr><td colspan="2" rowspan="3">Jurassic (Hettangian - Oxfordian) to Lr. Cretaceous (Valanginian)</td><td rowspan="4">Lagundsari Group</td><td rowspan="2">Chukku Fm</td><td colspan="2">Upper Shale Member</td></tr>
<tr><td colspan="2">Basal Limestone</td></tr>
<tr><td colspan="3">Giumal Fomation</td></tr>
<tr><td colspan="2" rowspan="2"></td><td colspan="3">Spiti Shale Formation</td></tr>
<tr><td colspan="2">Tagling Limestone</td><td rowspan="2">Kioto Limestone Formation</td><td rowspan="11">L I L A N G G R O U P</td></tr>
<tr><td rowspan="6">Upper Triassic</td><td>Rhaetian</td><td colspan="2">Megalodon Limestone (Para stage)</td></tr>
<tr><td rowspan="3">Norian</td><td colspan="2">Quartzite Series</td><td>Nunuluka Formation</td></tr>
<tr><td colspan="2">Monotis shales</td><td>Alaror Formation</td></tr>
<tr><td colspan="2">Coral Limestone</td><td>Hangrang Formation</td></tr>
<tr><td rowspan="2">Carnian</td><td colspan="2">Juavites Beds
Tropites Beds
Grey Beds</td><td>Sanglung Formation</td></tr>
<tr><td colspan="2">Halobia Beds</td><td rowspan="2">Chomule Formation</td></tr>
<tr><td rowspan="3">Middle Triassic</td><td rowspan="2">Ladinian</td><td colspan="2">Daonella Limestone</td></tr>
<tr><td colspan="2">Daonella Shales</td><td>Kaga Formation</td></tr>
<tr><td>Anisian</td><td colspan="2">Upper Muschelkalk
Lower Muschelkalk
Nodular Limestone
Basal Muschelkalk</td><td rowspan="2">Mikin Formation</td></tr>
<tr><td>Lower Triassic</td><td>Scythian</td><td colspan="2">Hedenstroemia Beds
Meekoceras Zone
Ophiceras Zone
Otoceras Zone</td></tr>
<tr><td>Permian</td><td colspan="5">PRODUCTS SHALE - Kuling Group (Gungri Formation)</td></tr>
</table>

Table 14. Lilang Supergroup, Spiti, Himachal Pradesh

Age		Groups	Formation (Fm)	Member (M)
	Upper Triassic	KIOTO GROUP		
Lilang Supergroup		Nimuloksa Group	Numuluka Fm Alaror Fm Hangrang Fm Rangrik Fm	 Upper Member Lower Member
		Sanglung Group	Rongtong Fm Rama Fm Chomule Fm	Upper Member Middle Member Lower Member
	Middle Triassic	Tamba Kurkur Group	Kaga Fm Mikin Fm	 H. Muschelkalk M Niti Limestone M
	Lower Triassic			Limestone & Shale M Lower Limestone M

Table 15. Mesozoic of Kutch, Gujarat (Thickness in Meters)

Age		Formation	Beds
Cretaceous	Post Aptian	Umia (1000)	Bhuj Beds (Umia Plant Beds): sandstone and shale
	Aptian		Ukra Beds : marine calcareous shales
	Upper Neocomian		Umia Beds Barren sandstones & shales (unfossiliferous)
	Valanginian		Trigonia Beds Barren sandstones
	Upper Tithonian		Umia ammonite bed
Jurassic	Middle Tihonian Lower Katrol Middle Kimmeridgian	Katrol (300)	Upper Katrol Shales Gajansar Beds Upper Katrol (barren) sandstone Middle Katrol: red sandstone Lower Katrol: sandstone, shales, marls Kanthkot sandstone (Bimammantum Zone)
	Upper Oxfordian Upper to Lower Oxfordian Middle Callovian	Chari (360)	Dhosa oolite: green and brown oolites (Transversarium zone) Athleta beds: marls and gypseous shales Anceps beds: marls and gypseous shales Rehmanni beds: yellow limestone Macrocephalus beds: shales with calcareous bands, with golden oolite-diadematus zone in upper part
	Lower Callovian Lower Callovian to Bathonian	Pachcham (300)	Pachcham coral bed Pachcham shell limestone Pachcham basal beds (Kuar Bet Beds)

Table 16. Stratigraphy of Western Rajasthan

Age	Formation (Fm)		
	Barmer Basin	Jaisalmer Basin	Bikaner-Nagaur Basin
Recent	Dune sands, alluvium, gravel/Trap wash	Sand dunes	Sand dunes
Pleistocene to sub-Recent	Uttarlai Fm	Shumar Fm Bap Boulder Bed Fm	Mar Fm
Eocene	Kapurda Fm Acid & Basic dykes & skills Mathaji ka Dungar Fm Akli Fm	Bandah Fm Khuiala Fm	Jogira Fm
Palecene to Eocene	Barmer Fm Fatehgarh Fm	Sanu Fm	Marh Fm Palana Fm (not exposed)
Cretaceous	Sarnu Fm	Parh Fm (not exposed) Goru Fm (not exposed) Habur Fm Pariwar Fm	
Jurasic	Jaisalmer Fm Lathi Fm	Bhadasar Fm Baisakhi Fm Jaisalmer Fm Lathi Fm	
Permian to Triassic		Bhuana Fm (not exposed)	
Permian			Badhura Fm
Carboniferous to Permian			Bap Boulder Bed Fm (?)
Proterozoic to Cambrian	Birmania Fm	Birmania Fm	Nagaur Fm Bilara/Hansera Fm
Proterozoic (Precambrian)	RANDHA FORMATION		Jodhpur Fm
	MALANI SUITE BASEMENT COMPLEX		Basement Granite & Rhyolite

Table 17. Malla Johar Supergroup, Kumaon, Uttarakhand

Group	Formation (Fm)	Lithology	Age
Sancha Malla	Balcha Dhura Fm {90}	Basic volcanic rocks, shale, Radiolarian cherts	? Post Cretaceous
	Jhangu Fm 400}	Greywacke, shale, sandstone foram, limestone, chert	Upper Cretaceous
	Giumal Sandstone Fm {400}	Glauconite sandstone, silt, Radiolarians	Lower Cretaceous
	Spiti Shale Fm {250}	Shales, sits, nodules, ammonites	Upper Jurassic to Lower Cretaceous
Rawali-Bagar	Ferruginous oolitic Fm {10}	Ferruginous oolitic limestone with abundant ammonites	Middle Jurassic

Table 18. Stratigraphy of Trichinopoly Area

Age	Group	Formation (Thickness in m)	Lithology
Early Paleocene	Niniyur	Niniyur (60)	
Uppermost Cretaceous	Ariyalur	Kallamedu (1)	Sandstone and variegated clays
		Ottakovil (60)	Calcareous sandstones and siltstones
		Kallankurichi (40)	Limestones & Marls
		Silakudi (500)	Sandstone & some limestone
Upper Cretaceous	Trichinopoly	Anaipadi (100)	Sandstone, siltstone and concretionary bodies
		Kulakkalnattam (50)	Limestone, sandstone, siltstones and concretions of arenaceous limestone
Lower to Upper Cretaceous	Uttatur	Karai Shale (400)	
Lower Cretaceous		Maravattur Clay (400) Kallakkudi limestone (Dalmiapuram) (75)	
		Grey shale (50)	Dark grey sale, pyrites, woody remains
		Boulder Conglomerate (5)	
	Upper Gondwana	Terani plant beds (30)	
Archean	Crystallines		
Notes: 1. Except Boulder Conglomerate and Grey Shales, all other Formation are richly fossiliferous. 2. See Trichinopoly area.			

Table 19. Cenozoic Era

Subera	Period	Epoch	Age (Ma)
Quaternary	Neogene	Holocene (Recent) Pleistocene	0.0116 1.64
Tertiary		Pliocene Miocene	5.20 23.30
	Paleogene	Oligocene Eocene Paleocene	35.40 56.50 65.50

Table 20. Cenozoic Stratigraphy of Gujarat

Age			Mainland	Saurashtra	Kutch	
			Surface	Surface	Surface	Lakhpat Well - 1 Subsurface
QUATERN TERTIARY	NEOGENE	Holocene	Narbada Alluvium	Soils, alluvium, gravel, sands	Alluvium, Ranns	
		Pleistocene	Absent	Chaya Fm Miliolite	Miliolite	Miliolite Limestone
		Pliocene	Absent Jagadia Fm	Dwarka Fm Piraim Beds	Kankawati Sr –	Vinjhan Stg
		Miocene	Kand Fm Babaguru Fm	Kuda Sat Bumbli Cong	Khari Sr	Aida Stg Waior Stg Ramania
	PALEOGENE	Oligocene	Absent	Ratanpur Clay	Bermoti Sr	
		Eocene	Dinod Fm	Laterite	Berwali Sr Madh Sr	Babia Stg Kakdi Stg
		Paleocene (?)	Vagadkol Fm			
Mesozoic		Cretaceous	Deccan Traps		XXXXXX	

Fm= Formation; Stg = Stage; Sr = Series; Sst = Sandstone; Cong = Conglomerate

Table 21. Stratigraphy of Andaman-Nicobar Islands

Group	Formation (Fm)	Lithology	Age
Archipelago Group	Shell Limestone, beach Sand	Coral rag and sand	Sub-recent to Recent
----------------------- Unconformity -----------------------			
	Neil West Coast Fm	Bio-calcarenite	Late Pliocene to Pleistocene
	Guitar Fm	Calcareous mudstone	Early to Late Pliocene
	Sawai Bay Fm	Calcareous mudstone	Late Miocene to Early Pliocene
	Long Fm	Calcareous siltstone and silty mudstone	Middle to Late Miocene
	Inglis Fm	Calcareous chalk	Early Miocene
	Strait Fm (Round Fm)	Shell sandstone, limestone & conglomerates	Early to Middle Miocene
----------------------- Unconformity -----------------------			
Upper Cretaceous to Palaeogene Group	Port Blair Fm	Conglomerate, grit and greywacke	Middle Eocene to Oligocene
	Baratang Fm	Shale-greywacke	Late Cretaceous to Eocene
	Port-Meadow Fm	Radiolarian chert, quartzite	Late Cretaceous

Table 22. Stratigraphy of Sub-Himalayan Region

<table>
<tr><th>Age</th><th>Sindh-Salt Range (Paki-stan)</th><th>Potwar-Jammu</th><th colspan="2">Himachal-Kumaon</th><th>Arunachal Pradesh</th><th>Assam-Indo Myan-mar Range</th><th></th><th>Upper Sindhu Valley (Jammu & Kashmir)</th></tr>
<tr><td>Early Miocene</td><td>Gaj Fm</td><td>Murree - Fm</td><td rowspan="4">Simur Gr</td><td>Kasauli Fm</td><td rowspan="2"></td><td>Surma Gr</td><td>Bokabil Fm
Bhuban Fm</td><td rowspan="3">Kailas Conglo-merate (Kargil Fm)
Indus Fm</td></tr>
<tr><td>Oligo-cene</td><td>Nari Fm</td><td rowspan="2">Chharat Lime-stone Fm</td><td>Dasghai Fm</td><td rowspan="2">Barail Gr</td><td>Renji Fm</td></tr>
<tr><td rowspan="2">Eocene</td><td>Kirthar Fm</td><td rowspan="2">Subathu Fm</td><td rowspan="2">Yinkiong Gr</td><td>Jenam Fm
Laisong Fm</td></tr>
<tr><td>Laki Fm</td><td rowspan="2">Hill Lime-stone</td><td rowspan="2">Jaintia / Disang Fm</td><td rowspan="2">Kopili Fm
Sylhet
Lakadong (Langpur)
Theria</td><td rowspan="2">Indus Fm</td></tr>
<tr><td>Paleocene</td><td>Ranikot Fm</td><td colspan="2">Bansi-Kakara Fm</td><td></td></tr>
<tr><td>Permian</td><td></td><td></td><td colspan="2"></td><td>Abor Volcanic Formation</td><td></td><td></td><td></td></tr>
</table>

Fm=Formation, Gr=Group

Table 23. Stratigraphy of NE India

Age (Probable)	Group	Sub group	Garo Hills		Khasi & Jaintia Hills		Mikir Hills & Dhansiri Valley	Upper Assam Valley	Naga Hills Fold Belt	Surma Valley & Tripura Fold Belt
Holcene			Alluvium		Alluvium		Alluvium	Alluvium	Alluvium	Alluvium
Pleistocene to Pliocene		Dihing	Dalu Fm Rangapani Fm		....		Disama Fm	Dhekiajuli Fm		Dhing Fm
Pliocene to Miocene		Dupitila	Bikona Fm		Dupitila Fm		Rangpahar Fm	Namsang Fm	Namsang Fm Rangpahar Fm in SW part	Dupitila Fm
	Surma	Tipam	?		Tipam Fm		Tipam Fm	Up. Sst Fm Girujan Fm Tipam Sst Fm	Girujan Fm Tipam Sst Fm	Tipam Fm
		Bokabil	Angartoli Fm		Bokabil Fm		Bokabil Fm	Bokabil Fm	Bokabil Fm	Bokabil Fm
		Bhuban	Boldamgiri Fm		Up Bhuban M, Mid Bhuban M Lr Bhuban M					
	------------ Unconformity ------------									
Oligocene		Barail	Khera-para Fm	Inolgiri M	Renji Fm Jenam Fm		Renji Fm Jenam Fm		Tikak Parbat Fm Baragolai Fm	Renji Fm Jenam Fm
				Darik M	Laisong Fm		Laisong Fm	Arenaceous facies	Naogaon Fm	Laisong Fm
Eocene	Jaintia	Disang Fm	Rewak Fm		Kopili Fm		Kopili Fm	Kopili Fm		
Paleocene			Sylhet subgroup	Siju Fm Tura Fm	Sylhet subgroup	Prang M Narpuh Sst M Umlatdoh Lst M Lakadong Sst M Lakadong Lst M	Sylhet Lst Fm Tura Fm	Sylhet Lst Fm Tura Fm Basal Sst	Disang Fm	Disang Fm
				Langarin M	Theria Fm	Therria Sst/ Cherra M Therria Lst M				
Late Cretaceous	Khasil		.....Unconformity.....			Langpar Fm	..Unconformity..			
			Mahadek Fm							
			Gurnaghat Fm							
Early Cretaceous to Permian			Sylhat Trap Gondwana Gr			Sylhat Trap	Mikir Hills Trap Gondwana Gr			
Precambrian	------------ METAMORPHIC COMPLEX ------------									

Fm = Formation, M=Member, Up = Upper, Mid = Middle, Lr = Lower, Sst = Sandstone, Lst = Limestone, sub = subsurface

Table 24. Karewa Group, Kashmir Valley

Age	Formation (FM)	Member (M)	Lithology
Pleistocene	Dilpur Fm Nagum Fm	.... Shuplyan M Pampur M Krungus M	Loam sediments Fine to coarse ochre Sandy clay ochre Marl, gravel, ochre
Pliocene	Hirpur Fm (sporadically exposed in the NE part)	Methawoin Rembiara M Dubjan M	Plastic sandy clay, fine to coarse sand, conglomerate, lignite and lignite clay

Table 25. Stratigraphy of Kerala

Age	Supergroup	Group	Formation (Fm)	Lithology
Quaternary	Malabar Supergroup		Vembanad Fm	Beach sands, alluvium, red (ten), sands, peat beds with semi-carbonised wood, calcareous clays, etc.
---------------- Unconformity (marked by laterite) ----------------				
Tertiary (Miocene)	Malabar Supergroup	Warkalli Group	Ambalapuzha Fm Quilon Fm (including Vaikon Fm) Mayyanad Fm	Sandstones, clays and lignites Limestones, marlsm calcareous clays and sands Sandstones, coarse gravely sands, clays, lignites and carbonaceous clays
---------------- Unconformity ----------------				
Late Mesozoic			Karruchal Fm	Hard, compact, ferruginous, gritty sandstones, conglomeratic near base with clay interbeds and arkoses in the upper part
---------------- Unconformity (marked by bauxite) ----------------				
Archaean	Gneisses, granulites and schists			

Table 26. Stratigraphy of Painkhanda (Niti area), Uttarakhand

Age			Formation (Fm)	Member (M)	Lithology
Era	Period	Sub-division			
Mesozoic	Jurassic	Tithonian	Kiogad Fm	Spiti M	Grey to black shales with calcareous nodules and rich in ammonites
		Oxfordian Kimmeridgian		Chhojan M	Oolitic ferruginous, arenaceous limestone with Belemnites
		Callovian		Lapthal M	Thinnly bedded limestone full of Lumachella
	Triassic	Norian	Shal Shal Fm	Kioto M	Limestone, arenaceous limestone and quartzite with rare fossils of Megalodon, Halorites, etc.
				Rimkhim M	Limestone and shale with Halobia, etc. fossils
				Hoti M	Limestone and shale alternation
				Niti M	Massive nodular limestones
				Rambakot M	Limestone and shale alternation with otoceras and ophiceras fossils
Palaeozoic	Upper Permian		Kuling Fm		Black shale with nodules and fine grained sandstone at the base, In the middle limestone lenses contain fossils of Spiriferella rajah and Paramarginifera himalayensis

Table 27. Quaternary sediments of Kali Valley, Kumanon, Uttarakhand

Age	Litho unit	Lithology
Holocene (Recent)	Kali Recent Alluvium	Sand, silt, gravel. Pebble, cobble and boulder
	Recent Alluvim/ colluvial fan deposits	Associated rock fragments in sandy/silty matrix
	Recent Glacial deposits	Assorted rock fragments embeded in rock flour matrix
Holocene (Recent)	Kali Younger Terrace Alluvium	Sand, silt, clay, gravel. Pebble, cobble and boulder size clasts
	Outwash Deposits	Gravel zone, beds of silt/clay
	Solifluction Deposits	Fragments of country rocks with silt/clay blanket

	Younger Glacial Deposits	Rock fragments in rock flour matrix
Middle to Late Pleistocene?	Kali Older Terrace Alluvium	Pebble to boulder size clasts in sandy/silty matrix
	Valley Train Deposit/ Glacio-fluvial Deposit	Cobbles with few gravel, pebble in stratified sediments
	Glaciolacustrine/ Varve Deposit	Thinly bedded, stratified, silt/clay couplets
	Older Glacial Deposit	Assorted rock fragments in oxidised matrix
-------------------------- Unconformity --------------------------		
Pre-Quaternary rocks of Garhwal Group. Central Crystalline and Tethyan Sedimentaries		

Table 28. Siwalik Group

Age	Division	Formation (Fm)	Lithology
Pleistocene	Upper (2300)	Boulder Conglo-merate Fm	Predominantly massive conglomerate with clays as matrix and minor sandstone and claystone
		Pinjor Fm	Sandstone, clay and conglomerate alternation
		Tatrot Fm	
Pliocene	Middle (1400 to 2000)	Dhok Pathan (Fm)	Massive sandstone with minor conglomerate and local variegated claystone
		Nagri Fm	Predominantly sandstone and red clays alternation with subordinate claystone and locally conglomerate
Miocene	Lower (1600)	Chinji Fm	Alternation of sporadically pebbly sandstone, calcareous cement and prominent chocolate and maroon claystone in the middle part
		Kamlial Fm	Red and massive claystone with intercalations of medium to fine grained sandstone
Note: Clays and claystones are generally red, orange, buff or brown in colour			

Table 29. Basins Producing Hydrocarbons

Basins	Petroleum source Sediments	
Bombay offshore	Paleocene-Early Eocene sediments in Surat Depression (Panna Fm)	Mainly Miocene carbonates
Cambay	Lower Eocene Cambay Shale	Middle Eocene Sandstones of Ankaleshwar and Kabal Formations
Upper Assam	Eocene shales and Coal shale sediments of Barail (Oligocene)	Barail Sandstone (Oligocene)- Tipam Sandstone (Miocene)

Krishna Godavari	Early Cretaceous shales in deeper part of the basin	Paleocene-Eocene Sandstone (Pasariapudi Formation), Ravva Sandstones (Miocene)
Cauvery	Early Cretaceous shale Andimadam Formation	Sandstones of Bhuvanagiri and Nannilam Formations (Upper Cretaceous), Kamalapuram Formation (Paleocene-Eocene), Niravai Sandstone (Oligocene)
Note: Cambay and Upper Assam Basins have structural Traps, while other Basins have structural cum stratigraphy Traps.		

Table 30. Godavari Supergroup (Age : Proterozoic) (Thickness in Meters)

Pakhal or Mallampalli Belt area		Albaka Belt area	
GONDWANA SUPERGROUP {Late Paleozoic - Mesozoic}			
Angular Unconformity		Fault	
Group	Formation	Group	Formation
Sullavai Group {870}	Venkatapur Sandstone Mancharial Quartzite Ramagiri Fm	Usur Group {1375}	Delman Sandstone Doli Sandstone Nambi Fm
------ Angular uncoformity ------		------ Disconformity ------	
Penganga Group {2700}	Sat Nala Shale Chanda Limestone / Puntur Limestone Pranhita Sandstone Takalapalli Arkose	Penganga Group {1000}	Albaka / Pamanuru {Chelimala} Sandstone with basal Utlapalli Conglomerate
------ Disconformity ------		------ Disconformity ------	
Mulug Group {2830}	Laknavaram Shale Pattipalli Quartzite Enchencheruvu Dolomite Polaram Shale Jakaram Arkose	Venkata-puram Group {2250}	Tippapuram Shale Somadevara Quartzite
------ Disconformity ------		------ Disconformity ------	
Mallam-palli Group {825}	Pandikunta Shale / Karlai Shale Gunjeda Dolomite / Bayyaram Quartzite Bolapalli Fm		Cherala Fm/Somanpalli Gr
------------ Non-conformity / Angular unconformity ------------			
------ Crystalline Basement ------		------ Crystalline Basement ------	
(Dharwar craton)		(Bastar craton)	

Fm = Formation

Table 31. Impotant Geological Units (Formations, etc.) of India

Age	Geological units
Archean	Rocks of the Dharwar Craton: see Dharwar Craton & Tables 2 & 3
	Rocks of the Singhbum: see Singhbhum Craton and Table 5
	Rocks of the Bundelkhand Craton: see Bundelkhand Craton
Archean & Paleoproterozoic	Rocks of the Bastar Craton: see Bastar Craton & Table 4
	Rocks of the Aravalli Craton:
Archean	Mewar Gneiss, Bhilwara Supergroup and Mineralised Supracrustal Belts
Paleoproterozoic	Aravalli Supergroup: see Aravalli Supergroup & Table 6
Neoproterozoic	Delhi Supergroup: see Delhi Supergroup & Table 6
Neoproterozoic to Holocene (Recent)	Rocks of the Marwar Craton: see Tables 6 & 16
Proterozoic	Rocks of the Eastern Ghats Moible Belt, Pandyan Mobile Belt and Satpura Mobile Belt: see each Belt
	Cuddapah Supergroup and Kurnool Group, Chhattisgarh Supergroup, Godavari Supergroup and Vindhyan Supergroup. See each of these
Precambrian	Precambrian of Himalaya: see Precambrian of Himalaya
Cambrian	Dogra Slates, Lolab, Batal, Kunzum La and Tal Formations of Jammu & Kashmir: see each of these
Ordovician	Thango Formation of Himachal Pradesh: see this
Ordovician & Silurian	Takche (Manchap) Formation of Himachal Pradesh, Variegated Formation, Shiala Formation, Garbyang Formation of Garhwal-Kumaon, Uttarakhand see each of these
Devonian	Much Quartzite of Jammu & Kashmir, Himachal Pradesh and Uttarakhand
Carbonifrous	Tannaki Conglomerate, Syringothyris Limestone, Fenestella Shale and Agglomeratic Slate and Sasar Brangsa/Shaksgam Formation of Jammu and Kashmir, Kanwar Group (Lipak & Po Foramtions) of Himachal Pradesh and Uttarakhand. Ganmachidam Formation of Himachal Pradesh
Palaeozoic (probably Permian)	Tandi Group, Himachal Pradesh
Permian	Nishatbagh Formation, Panjal Volcanics, Mamal Formation, Zewan Formation of Jammu & Kashmir, Kuling Formation of Painkhanda, Uttarakhand. Gechang Formation and Gungeri Formation of Himachal Pradesh. Seti Khola Formation of Bhutan, Miri Formation, Yamne Formation of Arunachal Pradesh
Permo-Carboniferous	Umaria Marine Bed, Madhya Pradesh

Age	Geological units
Carboniferous to Cretaceous	Gondwan Supergroup. See Gondwana Supergroup and Table 12.
Triassic	Khunamuh Formation of Kashmir, Lilang Group, Spiti Kalapani Limestone Chocolate Series and Ktui Shales of Byans, Uttarakhand. Shal Shal Formation of Painkhanda, Uttarakhand
Permian to Cretaceous	Exotic Blocks of Malla Johar, Uttarakhand
Jurassic	Kiogad Formation Painkhanda, Uttarakhand
Jurassic to Cretaceous	Pachcham, Chari, Katrol and Umia Formations of Kutch, Gujarat. Malla Johar Supergroup, Uttarakhand. see Table 17.
Proterozoic to Holocene (Recent)	See Table 16 - Stratigraphy of Western Rajasthan
Cretaceous to Paleocene	See Table 18 - Stratigraphy of Trichinopoly area.
Createaceous	Bagh Formation and Lameta Formation of Madhya Pradesh Porlob Group of Andaman Nicobar Islands
Cretaceous to Eocene	Serpentine Group of Andaman Nicobar Islands Hydrocarbon producing Basins of India. see Table 29.
Late Cretaceous to Early Paleocene	The Deccan Volcanic Province of the Deccan Traps, alongwith Infra-trappean and Inter-trappean Beds
Paleocene to Holocene	See Table 20 - Cenozoic Formation of Gujarat
Cretaceous to Holocene (Recent)	See Table 21 - Stratigraphy of Andaman Nicobar Islands
Paleocene to Miocene	See Table 22 - Stratigraphy of sub-Himalayan Region
Precambrian to Holocene (Recent)	See Table 23 - Stratigraphy of NE India
Piocene to Pleistocene	See Karewa Group and Table 24
Eocene	Pash Kyum Formation of Kargil, Kashmir
Miocene	Baripada Beds, Orissa Rajahmundry Formation, Andhra Pradesh
Miocene to Pliocene	Cuddalore Sandstone Froamtion of Tamil Nadu
Mesozoic to Quaternary	Stratigraphy of Kerala. See Table 25
Quaternary	See Quaternary Deposits of Himachal Pradesh, see Table 32 Quaternary of Kali Valley, Uttarakhand, see Table 27
Miocene to Pleistocene	See Siwalik Group and Table 25
Pleistocene to Holocene	Indo-Gangatic alluvium and Desert sands. See Indo-Gangetic Plain

All the Formations, Tables, etc. mentioned below are given in this book please see

Table 32. Purana or The Proterozoic Sedimentary Basins

Craton	**Paleoproterozoic Basins**	**Meso-Neoproterozoic Basins**
Bundelkhand Craton	Bijawar and Sonrai Basins and Harda Inlier Gwarior Basin	Vindhyan Basin
Bastar Craton	Abujhmar Basin	Chhattisgarh Basin Khariar Basin Ampani Basin Keskal, Singanpur and Chedrapal outliers Indravati Basin Sabri Basin
Between Bastar and Dharwar Craton		Pranhita- Godavari Basin
Dharwar Craton	Papaghni sub-Basin of Cuddapah Basin	Cuddapah Basin Kaladgi Basin Bhima Basin

See Purana Basins and each Basin separately also.

Table 33. Chattisgarh Supergroup, Bastar Craton

Group	Formation and lithology
Raipur Group {1900}	Maniari Shale: Shale, dolomite and gypsum
	Hiri Dolomite: Grey dolomite
	Tarenga Shale: Shale, shale-chert beds and limestone
	Chandi Limestone: Stromatolitic dolomite, limestone, sandstone, shale
	Gunderdehi Shale: Shale, limestone, ignimbrite
	Charmuria Limestone: Phosphatic limestone/dolomite, hale, chert
	Bijepar Shale : shale with sandy interbeds
Chandraput Group {400}	Kansapathar Sandstone : Galuconitic sandstone
	Chapordadih Shale: Shale, arenite
	Lohardih Conglomerate: arkose, arenite, conglomerate
---------------------- unconformity----------------------	
Singhora Group {400}	Chhuipali Shale: Shale, chert, limestone, dolomite, siltstone
	Bhalukona Sandstone: Sandstone, siltstone, a little shale
	Saraipali Shale : Shale, siltstone, limestone, porcellanite, felsic tuff
	Rehatikhol Conglomerate: Arenite, arkose, conglomerate
---------------------- Nonconformity ----------------------	
Crystalline Basement of Bastar Craton	

Thickness in Metres

Table 34. Stratigraphy of Bijawar Group, M.P. Bundelkhand Craton

Group	**Sub-group**	**Formation**
Bijawar Group	Gangu subgroup	Karri Sandstone Hirapur Phosphorite
	Moli subgroup	Malehra Chert Breccia Pukhra Sandstone Dargawan Sill Bajno Dolomite Bhusor Basalt Kawar Conglomerate

Table 35. Stratigraphy of Pairi Group

Group	**Formations**
Pairi Group	Ling Dongi Sandstone
	Tarjhar Shale
	Galighat Sandstone
Disconformity	
	Neor Sandstone
	Klharighat Limestone
	Devdahra Sandstone
Unconformity	
Crystallines of Bastar Craton	

Table 36. Stratigraphy of Indravati Basin, M.P., Bastar Craton

Formaton (Gm)	**Lithology**
Jagdalpur Fm	Shale and stromatolitic dolomite
Kanger Limestone Fm	Limestone and shale
Cherakur Fm	Shale and siltstone with sandstone, chert pebble conglomerate
Tiratgarh Fm	Quartz arnite with shale, subarkose and quatz pebble conglomerate
---------------------- Unconformity----------------------	
------------------ Granite, Gneisses and supracrustal enclaves ------------------	

Table 37. Stratigraphy of Kaladgi Basin, Proterozoic Age, Dharwar Craton (Thickness in Meter)

Group (Gr)	Subgroup	Formation	Lithology
Badami Gr (285)		Katageri Limestone Kerur Arenite	Limestone, shale Conglomerate, arenite, shale
---------------------- Angular Unconformity ----------------------			
---------------------- Intrusives-Quartz veins, pegmatites, dolerite dykes ----------------------			
Bagalkot G	Simikeri Subgroup (1150)	Hosakatti Argillite Arlikatti Dolomite Niralkeri Breccia Kundargi Quartzite	Artillite Dolomite, hematite bed Chert breccia Conglomerate, quartzite, argillite
	---------------------- Disconformity----------------------		
	Lokapur Subgroup (27500)	Yadahalli Argillite Muddapur Dolomite Chikshellikeri Limestone Yargatti Argillite Mahakut Breccia Saundatti Quartzite	Argillite Dolomite, limestone, argillite Limestone, shale Argillite, dolomite Chert breccia Conglomerate, quartzite, shale
Nonconformity and Angular Conformity			
Gneisses/Granites and Schist Belts of Dharwar Craton			

Table 38. Tertiary Stratigraphy of Kutch, Gujarat

Age	**Series**	**Stage**	**Formation**
Pliocene	Kankawati Sr		Sandhan Fm
Miocene	Khari Sr	Vinjhan Stg	Vinjhan Fm Khari Nadi Fm
Oligocene	Bermoti Sr	Aida Stg Waior Stg Ramania Stg	Maniyara Fort Fm
Middle Eocene	Berwali Sr	Babia Stg	Fulra Limestone Fm Harudi Fm
Early Eocene		Kakdi Stg	Nareda Fm
Paleocene	Madh Sr		Matanomadh Fm

Sr= Series; Stg = Stage; Fm = Formation

■■■■

V

LIST OF FOSSILES
(With Occurrences and Fossil Groups)

(Total number of fossil genera in this list are 1217 which belong to 54 different Fossil Group and 177 Formations)

Fm = Formation; Gr = Group; Ser = Series; Stg = Stage; SG=Supergroup;
TF=Tracefossil; PF=Palynofossil; NF = Nanofossil;
NA= The author was unable to find the Fossil Group
Note: The fossil which was at D54 has been deleted due to lack of conformity

Fossil No.	Fossil Name	Fossil Group	Formation and area	Formation Age	Formation Number
A 1	Abathomphalus mayaroensis	Foraminifera	Infra-trappean beds, Rajahmundry	Cretaceous	i 5-1b
A 2	Acanthodiscus	Ammonite	Spiti Formation/ Shale, Malla Johar Supergroup	Lr. Cretaceous	S 61
			Sancha Malla Group, Uttarakhand	Up. Jurassic to Cretaceous	S 5
A 3	Acanthorhynchia	Ammonite	Umia Stage, Kutch, Gujarat	Cretaceous	U 11
A 4	Acecularia	Algae	Kallakkudi Limestone Formation, Uttatur Gr, Trichinopoly	Cretaceous	K 11
A 5	Acicularia dyumatsenae		Niniyur Formation, Trichinopoly	Paleocne	N 23
A 6	Aceratherium	Rhinocerotidae	Siwalik Group, Lower Division	Miocene	S 51
A 7	Actinodon risinensis	Vertebrata	Gangamopteris Beds	Permian	G 5
A 8	Aequitriradites spinulosum	PF	Gondwana Supergroup, Upper Division	Mid. Triassic- Lr. Cretaceous	G 23b
A 9	Agittodontus sp.	Conodonta	Kunzun La Formation, Spiti	Cambrian	K 73a
A 10	Agriotherium	Carnivora	Siwalik Group, Upper Division	Pleistocene	S 51c

Fossil No.	Fossil Name	Fossil Group	Formation and area	Formation Age	Formation Number
A 11	Alectryonia	Lamellibranchia	Karai Shale Formation and Maravattur Clay Formation, Uttatur Gr, Trichinopoly	Cretaceous	K 26 & M28
			Ariyalur Group, Trichinopoly	Cretaceous	A 35
A 12	Alisporites	NA	Chari Formation and Katrol Formation, Kutch	Jurassic	C 16 K 38
			Umia Stage, Kutch, Gujarat	Cretaceous	U 11
A 13	Allocovioceras sp.	Ammonite	Kulakkalnattam Formation, Trichinopoly Gr, Trichinopoly	Cretaceous	K 69
A 14	Alloganthosuchus	Crocodylidae	Takli Formation, Nagpur	Cretaceous-Paleocene	T 4
A 15	Alpilina	Foraminifera	Hundri Formation, Jammu and Kashmir	Cretaceous	H 21
A 16	Altanicypris	Ostracoda	Inter-trappean beds, Mumbai	Cretaceous	i 8 -5a
A 17	Altanicypris bhatiai	Ostracoda	Infra-trappean (=Lameta) beds of Jabalpur, Madhya Pradesh	Cretaceous	i 5-3a
A 18	Altanicypris deccanensis	Ostracoda	Inter-trappean beds of Lakshmipur, Gujarat	Cretaceous	i 8-2d
A 19	Alveolina	Foraminifera	Dras Volcanics, Jammu and Kashmir	Cretaceous	D 49
A 20	Amblyochara concave	Algae	Infra-trappean (=Lameta) beds of Dongargaon and Pidura, Maharashtra	Cretaceous	i 5-4a&b
A 21	Amblypterus kashmirensis	Vertebrata	Gangamopteris Beds	Permian	G 5
A 22	Amesoneuron goeppert	Palm	Infra-trappean beds, Jabalpur	Cretaceous-Paleocene	i 5-3a
A 23	Ammobaculites	Foraminifera	Raghvapuram Formation	Lr. Cretaceous	R 2
A 24	Ammonia	Foraminifera	Quaternary sediments of Gujarat, Mainland	Quaternary	Q 5
A 25	Ammonia beccardi	Foraminifera	Kankawati Series, Kutch	Pliocene	K 21
			Khari Series, Kutch	Miocene	K 44

Fossil No.	Fossil Name	Fossil Group	Formation and area	Formation Age	Formation Number
A 26	Amorphognathus	Conodonta	Takche Formation, Spiti	Silurian-Ordovician	T 3
A 27	Amphicyon sp.	Carnivora	Siwalik Group, Lower Division	Miocene	S 51a
A 28	Ampulina bulbiformis	Gastropoda	Bagh Beds, Narmada Basin	Cretaceous	B 8
A 29	Anguidae indet	Reptilia	Inter-trappean beds of Naskal, Andhra Pradesh	Cretaceous	i 8-1c
A 30	Anodontophora gresbachi	Lamellibranchia	Kioto Member, Uttarakhand	Triassic	K 55
A 31	Anomalina rudio	Foraminifera	Inter-trappean beds of Rajahmundry, Andhra Pradesh	Cretaceous	i 8-1b
A 32	Anomocaraspis	Trilobita	Kunzun La Formation, Zanskar, J&K	Cambrian	K 73b
A 33	Anomocarella	Trilobita	Kunzun La Formation, Zanskar, J&K	Cambrian	K 73b
A 34	Anstrotrilina	Foraminifera	Kand Formation, Gujarat Mainland	Lr. Miocene	K 19
A 35	Anthacodon	Anthracothe-riidae	Siwalik Group, Middle Division	Pliocene	S 51b
A 36	Anthacotherium	Anthracothe-riidae	Siwalik Group, Middle Division	Pliocene	S 51b
A 37	Anthraconeila kansana	Lamellibranchia	Miri Formation, Arunachal Pradesh	Permian	M 54
A 38	Antiella	NA	Ordovician-Silurian rocks of Shans Abani Syncline, Handwar, J&K	Ordovician	O 7
A 39	Antilope cervicapra	Artiodactyla	Quaternary sediments of Central India	Up. Pleistoce-ne to Holocene	Q 4
A 40	Anura indet	Amphibian	Inter-trappean beds of Nagpur, Maharashtra	Cretaceous	I 8-5b
A 41	Apiduim indicum	Mammalia	Tethyan Basin, Himalaya	Ordovician	T 20
A 42	Apogon curvatus	Fish	Inter-trappean beds of Rangapur, Andhra Pradesh	Cretaceous	i 8-1d
A 43	Apogon spatulatus	Fish	Inter-trappean beds of Rangapur, Andhra Pradesh	Cretaceous	i 8-1d
A 44	Apogonidarum ovatus	Fish	Inter-trappean beds of Rangapur, Andhra Pradesh	Cretaceous	i 8-1d

Fossil No.	Fossil Name	Fossil Group	Formation and area	Formation Age	Formation Number
A 45	Aquilapollenites indicus	PF	Inter-trappean beds of Anjar, Gujarat	Cretaceous	i 8-2a
A 46	Aquilapollenites bengalensis	PF	Inter-trappean beds of Mohgaon-Kalan and Padwar/Ranipur, Madhya Pradesh	Cretaceous	i 8-4b,c
A 47	Araucariacites	PF	Chari Formation and Katrol Formation, Kutch	Jurassic	C 16, K 38
			Umia Formation, Kutch	Cretaceous	U 8
A 48	Araucarites	Coniferales	Umia Beds, Kutch, Gujarat	Cretaceous	U 7
A 49	Araucarites cutchensis	Coniferales	Raghvapuram Formation	Lr. Cretaceous	R 2
A 50	Arca granosa	Lamellibranchia	Chilka Lake deposit of estuarine shells, Orissa	Pleistocene	C 33
A 51	Archaeocalamites radiates	Plantae	Gund Formation (Syringothyris Lst.), Kashmir	Lr. Carboniferous	G 31
A 52	Archaeogosaurus ornatus	Vertebrata	Gangamopteris Beds	Permian	G 5
A 53	Archaeolithothamnium feddeni	Algae	Wadhwan Formation, Saurashtra, Gujarat	Cretaceous	W 1
A 54	Archaeolithothamnium lugeoni	Algae	Niniyur Formation, Tamil Nadu	Paleocene	N 23
A 55	Archaeosiggilaria sp.	Plantae (Lycopodiophyta-an extinct vesicular plant)	Gund Formation (Syringothyris Lst.), Kashmir	Lr. Carboniferous	G 31
A 56	Archaias angulatus	Foraminifera	Khari Series, Kutch	Miocene	K 44
A 57	Archidisckondon planifrons	Proboscidea	Pleistocene sediments of Lower Karewa, Pinjor of Upper Siwaliks	Pleistocene	P 37d
A 58	Arcotia indica	Gastropoda	Trichinopoly Gr. Trichinopoly	Cretaceous	T 44
A 59	Arenicolites	TF	Thango Formation, Kinnaur, Himachal Pradsh	Ordovician	T 23
A 60	Ariidae indet	Fish	Inter-trappean beds of Anjar and Lakshmipur, Gujarat	Cretaceous	i 8-2a,d
A 61	Asiatosuchus	Crocodylidae	Takli Formation, Nagpur	Cretaceous-Paleocene	T 4

Fossil No.	Fossil Name	Fossil Group	Formation and area	Formation Age	Formation Number
A 62	Aspidium	Algae	Takche Formation, Spiti	Silurian-Ordovician	T 3
A 63	Aspidoceras	Ammonite	Middle Katrol, Kutch	Jurassic	M 45
A 64	Aspidoceras cf. subwynnei	Ammonite	Spiti Formation	Jurassic	S 61
A 65	Aspidoceras iphiceroides	Ammonite	Spiti Formation	Jurassic	S 61
A 66	Assilina daviesi	Foraminifera	Subathu Formation, Uttarakhand	Eocene to Miocene	S 65
A 67	Assilina laminose	Foraminifera	Subathu Formation, Uttarakhand	Eocene to Miocene	S 65
A 68	Astarte-Turritella	Gastropoda	Bagh Beds, Narmada Basin	Cretaceous	B 8
A 69	Asteriacites	TF	Po Formation, Spiti-Kinnaur area	Carboniferous	P 38
A 70	Astropolithan	TF	Tethyan Basin, Himalaya	Late Precambrian	T 20
A 71	Ataxioceras kachhense	Cephalopoda	Kanthkot Sandstone, Kutch	Jurassic	K 23
A 72	Athyris	Brachiopoda	Umaria Marine Bed and Manendragarh Marine Bed, Madhya Pradesh	Permian	U 5 & M 22
A 73	Atlanticypris	Ostracoda	Inter-trappean beds of Asifabad, Nagpur, Mamoni and Gurmatkal	Cretaceous	i 8-1a,5b, 6a,3b
A 74	Atrypa	Brachiopoda	Takche Formation, Spiti	Silurian-Ordovician	T 3
			Saser Brangsa Formation	Carboniferous	S 13
A 75	Atrypa aff. Spinosa	Brachiopoda	Tethyan Basin, Himalaya	Silurian	T 20
A 76	Atrypa aspens	Brachiopoda	Lipak Formation	Carboniferous	L 24
			Muth Formation, Spiti	Devonian	M 66
A 77	Atrypa reticularis	Brachiopoda	Muth Formation, Malla Johar, Uttarakhand	Devonian	M 66
			Muth Formation, Byans, Uttarakhand	Devonian	M 66
A 78	Aulacoseira sp.	PF	Inter-trappean beds of Mohgaon-Kalan, Madhya Pradesh	Cretaceous	I 8-4b
A 79	Aulacosphinctes	Ammonite	Umia Stage, Kutch, Gujarat	Cretaceous	U 11

Fossil No.	Fossil Name	Fossil Group	Formation and area	Formation Age	Formation Number
A 80	Aulacosphinctes cf. latauchei	Ammonite	Spiti Formation	Jurassic	S 61
A 81	Aulacosphinc-toides	Ammonite	Spiti Member, Uttarakhand, Upper Katrol Sandstone, Kutch	jurassic	S 62, U 14
A 82	Aulacosphinc-toides aff. linoptychus	Ammonite	Spiti Formation	Jurassic	S 61
A 83	Aulacosphinc-toides aff. uhligi	Ammonite	Spiti Formation	Jurassic	S 61
A 84	Aulacosphinc-toides sparsicosta	Ammonite	Spiti Formation	Jurassic	S 61
A 85	Aulichnites	TF	Po Formation, Spiti-Kinnaur area	Carboniferous	P 38
A 86	Australiceras sp	Ammonite	Ukra Beds, Kutch	Cretaceous	U 2
A 87	Australiceras jacki	Ammonite	Coastal Gondwana Formations, Ramnat-hpuram, Tamil Nadu	Lr.Cretaceous	C 45
A 88	Austrotrillina howchini	Foraminifera	Khari Series, Kutch	Miocene	K 44
A 89	Aviculopecten	Lamellibranchia	Boulder Slate Form-ation, Yamne Formation, Miri Formation	Permian	B 71, Y 2, M 54
A 90	Aviculopecten cuneatus	Lamellibranchia	Agglomeratic Slate, J & K	Carboniferous-Permian	A 4
A 91	Aviculopecten mitechelli	Lamellibranchia	Umaria Marine Bed and Manendragarh Marine Bed	Permian	U 5 & M 22
A 92	Aviculopecten squamuliferous	Lamellibranchia	Umaria Marine Bed and Manendragarh Marine Bed	Permian	U 5 & M 22
A 93	Axis axis	NA	Quaternary sediments of central India	Up. Pleisto-cene to Holocene	Q 4
A 94	Azolla cretacea	PF	Inter-trappean beds of Naskal, Andhra Pradesh; Mohgaon-Kalan and Padwar/ Ranipur, Madhya Pradesh	Cretaceous	i 8-1c, 4bc
B 1	Bairdia krithe	Ostracoda	Inter-trappean beds of Rajahmundry, Andhra Pradesh	Cretaceous	i 8-1b

Fossil No.	Fossil Name	Fossil Group	Formation and area	Formation Age	Formation Number
B 2	Bairdia subdeltoidea	Ostracoda	Inter-trappean beds of Rajahmundry, Andhra Pradesh	Cretaceous	i 8-1b
B 3	Bairdoppilata talukdarai	Ostracoda	Niniyur Formation, Tamil Nadu	Paleocene	N 23
B 4	Belemnites	Ammonite	Macrocephalus beds, Kutch	Jurassic	M 2
			Chikkim Formation, Himachal Pradesh	Cretaceous	C 31
B 5	Belemnites gerardi		Chhojan Member, Uttarakhand	Jurassic	C 26
B 6	Belemnite guards		Karai Shale Formation and Maravattur Clay Formation	Cretaceous	K 26 & M 28
B 7	Belemnopsis sp	Ammonite	Gajansar Beds, Kutch	Jurassic	G 2
B 8	Belemnopsis gerardi	Ammonite	Spiti Formation, Spiti & Drumgam Formation, Kashmir	Up. Jurassic to Lr. Cretaceous	S 61 & D 50
B 9	Bellanus	Invertebrata (Cirripedia)	Kand Formation, Gujarat Mainland	Lr. Miocene	K 19
B 10	Bellerophon	Gastropoda	Shiala Formation, Uttarakhand	Ordovician	S 31
			Badhaura Formation, Bikaner-Nagaur Basin, Rajasthan	Permian	B 5
			Miri Formation, Arunachal Pradesh	Permian	M 54
B 11	Beltaneliformis sp. Cf. B. brunsae	NA	Krol Group, Nainital Syncline	Neoprote-rozoic	K 64
B 12	Berriasella sp.	Ammonite	Giumal Formation	Cretaceous	G 15
B 13	Bifungites	TF	Thango Formation, Kinnaur	Ordovician	T 23
B 14	Bipathodus stabilis	Conodonta	Syringothyris Limestone	Carboniferous	S 79
B 15	Bipliceras	Cephalopoda	Kanthkot Sandstone, Kutch	Jurassic	K 23
B 16	Birostriana	Lamellibranchia	Bagh Beds, Narmada Basin	Cretaceous	B 8
B 17	Bison	Bovidae	Siwalik Group, Upper Division	Pleistocene	S 51c
B 18	Bispathodus aculeatus	Conodonta	Muth Formation, Spiti	Devonian	M 66
B 19	Blanfordiceras	Ammonite	Spiti Member, Uttarakhand	jurassic	S 62

Fossil No.	Fossil Name	Fossil Group	Formation and area	Formation Age	Formation Number
B 20	Blanfordiceras aff. Wallichi	Ammonite	Spiti Formation, Spiti	Jurassic	S 61
B 21	Blanfordiceras cricki	Ammonite	Spiti Formation	Jurassic	S 61
B 22	Bochianites	Ammonite	Spiti Formation/ Shale, Malla Johar Supergroup	Lr. Cretaceous	S 61
B 23	Bolivina	Foraminifera	Broach Formation, Gujarat	Pliocene	B 72
			Miliolite / Porbandar Stone, Gujarat	Pleistocene	M 49
B 24	Bolivinoides decorates	Foraminifera	Sillakudi Formation, Ariyalur Gr	Cretaceous	S 37
B 25	Bos	Bovidae	Siwalik Group, Upper Division	Pleistocene	S 51c
B 26	Bos indicus		Pleistocene sediments of Kurnool, Ghod, Manjra, Pravara and Godavari	Pleistocene	P 37a
B 27	Bos namadicus		Pleistocene sediments of Central Narmada (upper group), Palmar	Pleistocene	P 37b
			Quaternary sediments of Tamil Nadu, Soyamali	Pleistocene	Q 9
B 28	Boselaphus	Carnivora	Quaternary sediments of Andhra Pradesh, flood plain of river Pampa	Pleistocene	Q 3
B 29	Boselephas ragocamelus	Carnivora	Quaternary sediments of Central India	Up. Pleisto-cene to Holocene	Q 4
B 30	Brachyphyllum	Coniferales	Umia Beds, Kutch, Gujarat	Cretaceous	U 7
B 31	Brachyphyllum expansum	Coniferales	Sivaganga Beds, Tamil Nadu	Lr. Cretaceous	S 49
B 32	Brachyphyllum morrisianum	Coniferales	Gondwana Suppergroup, Upper Division	Mid. Triassic-Lr. Cretaceous	G 23b
B 33	Brachyphyllum rhombicum	Coniferales	Inter-trappean beds, Rajmahal Basin	Cretaceous	i 8-9
B 34	Brachypotherium	Rhinocerotidae	Siwalik Group, Middle Division	Pliocene	S 51b

Fossil No.	Fossil Name	Fossil Group	Formation and area	Formation Age	Formation Number
B 35	Brachythyrina sp.	Rhinocerotidae	Kuling Formation, Uttarakhand	Permian	K 69
B 36	Brachythyris	Brachiopoda	Agglomeratic Slate, J & K	Carboniferous-Permian	A 4
B 37	Brahmapithecus	Primates	Nagri Formation	Pliocene	N 4
B 38	Brontotheres	Vertebrate	Subathu Group, Katakot, Metka, Moghala inliers, Rajauri District, Jammu	Eocene	S 65
B 39	Bruhathakayo-saurus matleyi	Dinosauria	Kallamedu Formation, Ariyalur Group, Trichinopoly	Cretaceous	K 12
B 40	Bubalus spp.	Artiodactyla	Pleistocene sediments of Kurnool, Ghod, Manjra, Pravara and Godavari	Pleistocene	P 37a
B 41	Buccanopsis wartha	Gastrpoda	Agglomeratic Slate, J & K	Carboniferous-Permian	A 4
B 42	Buriadia scwardii	Coniferales	Gondwana Supergroup, Lower Division	Permian to Triassic	G 23a
B 43	Burmasia	Lamellibranchia	Yamne Formation, Arunachal Pradesh	Permian	Y 2
B 44	Buxtonia kashmiricus	Brachiopoda	Agglomeratic Slate	Carboniferous-Permian	A 4
B 45	Bythocypris sp.	Ostracoda	Inter-trappean beds of Rajahmundry, Andhra Pradesh	Cretaceous	i 8-1b
C 1	Calceola sandatina	Coral	Tethyan Basin, Himalaya	Devonian	T 20
C 2	Callialasporites	PF	Chari Formation and Katrol Formation, Kutch	Jurassic	C 16 & K 38
			Umia Formation, Kutch	Cretaceous	U 8
			Raghvapuram Formation	Lr. Cretaceous	R 2
C 3	Callumispora	PF	Kuling Shale Formation, Uttarakhand	Permian	K 70
C 4	Calycoceras neuboldi	Ammonite	Karai Shale Formation and Maravattur Clay Formation	Cretaceous	K 26& M 28
C 5	Calymene	Trilobita	Shiala Formation, Uttarakhand	Ordovician	S 31

Fossil No.	Fossil Name	Fossil Group	Formation and area	Formation Age	Formation Number
C 6	Calzichondrilla crustans	Porifera	Kallakkudi Limestone Formation	Cretaceous	K 11
C 7	Camelopardalis	Giraffidae	Siwalik Group, Upper Division	Pleistocene	S 51c
C 8	Camerophonia cf. pillipsi	NA	Muth Formation, Byans Uttarakhand	Devonian	M 66
C 9	Candidarum sp.	Fishes	Inter-trappean beds of Lakshmipur, Gujarat	Cretaceous	i 8-2d
C 10	Candona	Ostracoda	Inter-trappean beds of Asifabad & Infra-trappean (=Lameta) beds of Dongargaon, Maharashtra	Cretaceous	i 8-1a & I 5-4a
C 11	Candona altanulaensis	Ostracoda	Infra-trappean (=Lameta) beds of Jabalpur, Madhya Pradesh	Cretaceous	i 5-3a
C 12	Candona amosi	Ostracoda	Inter-trappean beds of Anjar and Kora, Gujarat; Gurmatkal, Karnataka	Cretaceous	i 8-2a,c; 3b
C 13	Candona cf. C. hubeiensis	Ostracoda	Infra-trappean (=Lameta) beds of Jabalpur, Madhya Pradesh	Cretaceous	i 5-3a
C 14	Candona mysorephaseolus	Ostracoda	Inter-trappean beds of Gurmatkal, Karnataka	Cretaceous	i 8-3b
C 15	Candoniella sp.	Ostracoda	Infra-trappean (=Lameta) beds of Dongargaon, Maharashtra	Cretaceous	i 5-4a
C 16	Canis sp.	Carnivora	Pleistocene sediments of Kurnool, Ghod, Manjra, Pravara and Godavari	Pleistocene	P 37a
C 17	Cardiceramus parareberti	Lamellibranchia	Anaipadi Formation	Cretaceous	A 22
C 18	Cardita	Lamellibranchia	Kand Formation, Gujarat Mainland	Lr. Miocene	K 19
C 19	Cardita beaumonti	Lamellibranchia	Tirupati Formation	Lr. Cretaceous	T 34
C 20	Cardium giumalense	Lamellibranchia	Giumal Formation	Cretaceous	G 15
C 21	Carneithyris carnea (Terebratulids)	Brachiopoda	Sillakudi Formation, Ariyalur Gr.	Cretaceous	S 37

Fossil No.	Fossil Name	Fossil Group	Formation and area	Formation Age	Formation Number
C 22	Carteremys leithii	Reptilia (Tortoise)	Inter-trappean beds, Mumbai	Cretaceous	I 8-5a
C 23	Cataceramus	Lamellibranchia	Bagh Beds, Narmada Basin	Cretaceous	B 8
C 24	Cauveripollis	PF	Cuddalore Sandstone Formation, Neyveli Lignite Field, Tamil Nadu	Miocene	C 56
C 25	Cenomanella archiaciana	Gastropoda	Khalsi Formation, J & K	Cretaceous	K 42
C 26	Cervus	Anthracoth-eriidae	Pleistocene sediments of Kurnool, Ghod, Manjra, Pravara and Godavari	Pleistocene	P 37a
			Quaternary sediments of Andhra Pradesh, flood plain of river Pampa	Pleistocene	Q 3
			Siwalik Group, Upper Division	Pleistocene	S 51c
C 27	Cervus duvauceli		Quaternary sediments of Central India	Up. Pleistocene to Holocene	Q 4
C 28	Cervus sp.		Pleistocene sediments of Central Narmada (upper group), Palmar	Pleistocene	P 37b
C 29	Cervus unicolor		Quaternary sediments of central India	Up.Pleistocene to Holocene	Q 4
C 30	Cetaceans	Vertebrata	Subathu Group, Katakot, Metka, Moghala inliers, Rajauri District, Jammu	Eocene	S 65
C 31	Chanda ovatus	Fish	Inter-trappean beds of Rangapur, Andhra Pradesh	Cretaceous	i 8-1d
C 32	Chara sp.	Algae	Inter-trappean beds of Kora, Gujarat; Gurmatkal, Karnataka	Cretaceous	i 8-2c,3b
C 33	Chelae	Crustacea	Tirupati Formation	Lr. Cretaceous	T 34
C 34	Chelonia indet	Reptilia	Inter-trappean beds of Asifabad and Naskal, Andhra Pradesh; Upparhatti, Karnataka	Cretaceous	i 8-1a,c; 3c

Fossil No.	Fossil Name	Fossil Group	Formation and area	Formation Age	Formation Number
C 35	Cheloniceras aff. martini	Cephalopoda	Ukra Beds, Umia Formation, Kutch	Cretaceous	U 2
C 36	Chelydosaurus mahomensis	Vertebrata	Gangamopteris Beds	Permian	G 5
C 37	Chiastozygus litterarius	NF	Karai Shale Formation and Maravattur Clay Formation	Cretaceous	K 26& M 28
C 38	Chilotherium	Rhinocerotidae	Siwalik Group, Lower Division	Miocene	S 51a
C 39	Chlorophyta	Algae	Bagh Beds (Nimar Sandstone) Narmada Basin	Cretaceous	B 8
C 40	Choeromeryx	Anthracoth-eriidae	Siwalik Group, Upper Division	Pleistocene	S 51c
C 41	Chondrites sp.	TF	Subathu Formation & Dagshai Formation, Himachal Pradesh	Eocene-Miocene	S 65 & D 1
C 42	Chonets	Brachiopoda	Takche Formation, Spit & Syrigothyris Lst. Formation, Kashmir	Silurian-Ordovician Carboniferous	T 3 & S 79
C 43	Cibicides	Foraminifera	Niniyur Formation, Tamil Ndu	Paleocene	N 23
			Aida Stage, Kutch	Oligocene	A 6
			Broach Formation, Gujarat	Pliocene	B 72
C 44	Cibicides lobatulus		Miliolite / Porbandar Stone, Gujarat	Pleistocene	M 49
C 45	Cibicides mensilla		Infrara-trappean (=Lameta) beds of Rajahmundry, Andhra Pradesh	Cretaceous	i5-1b
C 46	Cidaris	Echinoidea	Kallakkudi Limestone Formation, Uttatur Gr, Trichinopoly	Cretaceous	K 11
C 47	Cladophlebis	Ferns (Filicales)		Cretaceous	U 7
C 48	Cladophlebis clentialata		Raghvapuram Formation		R 2
C49	Cladophlebis denticulate		Dharangadhara Group, Gujarat		D 28
C 50	Cladophlebis indica		Rajmahal Traps		R 9

Fossil No.	Fossil Name	Fossil Group	Formation and area	Formation Age	Formation Number
C 51	Cladophlebis lobata		Sivaganga Beds, Tamil Nadu		S 49
C 52	Cladophlebis sp		Gondwana Supergroup, coastal rocks	Cretaceous to Tertiary	G 23
C 53	Claria bioni	Fish	Khunamuh Formation, J & K	Permian-Triassic	K 51
C 54	Classopollis meditriangulus	PF	Gondwana Supergroup, Upper Division	Mid. Triassic-Lr. Cretaceous	G 23
C 55	Clupeidarum sp.	Fish	Inter-trappean beds of Rangapur, Andhra Pradesh	Cretaceous	i 8-1d
C 56	Clydagnathus gilwerensis	Conodonta	Syringothyris Lst. Formation, Kashmir	Carboniferous	S 79
			Muth Formation, Kashmir, Kishtwar, Ladakh and Kumaon	Devonian	M 66
			Tethyan Basin, Himalaya	Carboniferous	T 20
C 57	Clydagnathus unicornis		Muth Formation, Kashmir, Kishtwar, Ladakh and Kumaon	Devonian	M 66
			Tethyan Basin, Himalaya	Carboniferous	T 20
C 58	Clydagnatus cavusformis		Tethyan Basin, Himalaya	Devonian	T 20
C 59	Cochilichnus	TF	Tethyan Basin, Himalaya	Late Precambrian	T 20
			Po Formation, Spiti-Kinnaur area	Carboniferous	P 38
C 60	Codakia percrassa	Lamellibranchia	Niniyur Group/ Formation, Trichinopoly	Early Paleocene	N 23
C 61	Coelodonta	Rhinocerotidae	Siwalik Group, Up. Divi	Pleistocene	S 51c
C 62	Coenites	Coral	Takche Formation, Spiti	Silurian-Ordovician	T 3
C 63	Coilopoceras	Ammonite	Bagh Beds, Narmada Basin	Cretaceous	B 8
C 64	Collumnaefacta vulgaris	Algae (Stromatolite)	Tal Formation, Himachal Pradesh	Lr. Cambrian	T 5
C 65	Colombiceras	Ammonite	Ukra Beds, Umia Formation, Kutch	Cretaceous	U 2
C 66	Concavissimi-sporites sp.	PF	Raghvapuram Formation	Lr. Cretaceous	R 2

Fossil No.	Fossil Name	Fossil Group	Formation and area	Formation Age	Formation Number
C 67	Coniophis sp.	Reptilia	Inter-trappean beds of Naskal, Andhra Pradesh	Cretaceous	i 8-1c
C 68	Conohys sp.	Suidae	Siwalik Group, Lower Division	Miocene	S 51a
C 69	Contignisporites	PF	Chari Formation and Katrol Formation, Kutch	Jurassic	C 16& K 38
			Raghvapuram Formation	Lr. Cretaceous	R 2
C 70	Conularia cf. laevigata	Gastropoda	Badhaura Formation, Bikaner-Nagaur Basin, Rajasthan	Permian	B 5
C 71	Conularia laskari		Miri Formation, Arunachal Pradesh	Permian	M 54
C 72	Conus	Gastropoda	Kand Formation, Gujarat Mainland	Lr. Miocene	K 19
C 73	Copemys	Rodentia	Siwalik Group, Lower Division	Miocene	S 51a
C 74	Coptospora cauveriana	PF	Gondwana Supergroup, Up. Divi.	Mid. Triassic-Lr. Cretaceous	G 23b
C 75	Corbicula sp.	Lamellibranchia	Infra-trappean (=Lameta) beds of Dongargaon, Maharashtra	Cretaceous	i 5-4a
C 76	Corbula	Lamellibranchia	Pachcham shell Lst	Jurassic	P 4
			Pachcham basal beds, Pachcham Formation, Kutch		P 1
C 77	Cordaicarpus	Cordaitales	Gangamopteris Beds, Gondwana, Kashmir	Permian	G 5
C 78	Cordaicarpus zeilleri		Bahreli Formation, Arunachal Pradesh	Permian	B 10
C 79	Cordaites hislopi	Cordaitales	Gangamopteris Beds, Gondwana, Kashmir	Permian	G 5
C 80	Cordaites hislopi	PF	Nishatbagh Bed, J & K	Lr. Permian	N 25
C 81	Corisaccites	PF	Gondwana Supergroup, Lr. Divi	Permian to Triassic	G 23a
C 82	Cormohipparion	Equidae	Siwalik Group, Upper Division	Pleistocene	S 51c
C 83	Corongoceras	Ammonite	Spiti Member, Uttarakhand	jurassic	S 62
C 84	Costispirites	NA	Saser Brangsa Formation	Carboniferous	S 13

Fossil No.	Fossil Name	Fossil Group	Formation and area	Formation Age	Formation Number
C 85	Crassitriape-rturites gennatus	PF	Inter-trappean beds of Padwar/Ranipur, Madhya Pradesh	Cretaceous	i 8-4c
C 86	Crassostrea scidiformis	Invertebrata (Oyster)	Wadhwan Formation, Saurashtra, Gujarat	Cretaceous	W 1
C 87	Cremnoceramus	Lamellibranchia	Bagh Beds, Narmada Basin	Cretaceous	B 8
C 88	Crescentipollenites	PF	Barakar Formation, Pranhita-Godavari Valley	Lr. Permian	B 24
C 89	Cretirhynchia plicatiloides	Brachiopoda	Sillakudi Formation, Ariyalur Gr. Trichinopoly	Cretaceous	S 37
			Kulakkalnattam Formation, Trichin-opoly Gr, Trichinopoly		K 69
C 90	Crinoid spines	Lamellibranchia	Karai Shale Formation and Maravattur Clay Formation		K 26 M 28
C 91	Crocodylia indet	Crocodylidae	Inter-trappean beds of Asifabad, Naskal and Rangapur, Andhra Pradesh; Kisalpuri, Madhya Pradesh and Nagpur, Maharashtra	Cretaceous	I 8-1a,c,d; 4a;5b
C 92	Crocodylus palustris		Quaternary sediments of Andhra Pradesh, Gunglakamma and Munneru river basins	Pleistocene	Q 3
C 93	Crocuta	Carnivora	Siwalik Group, Mid.Divi.	Pliocene	S 51b
			Siwalik Group, Up.Divi.	Pleistocene	S 51c
C 94	Crossotrea gajensis	Invertebrata (Oyster)	Baripada Beds, Orissa	Miocene	B 26
C 95	Cruziana	TF	Kunzun La Formation, Spiti	Lr. Cambrian	K 73a
			Garbyang Formation, Uttarakhand	Cambrian	G 8
C 96	Cryassoretitriletes	PF	Cuddalore Sandstone Formation, Neyveli Lignite Field, Tamil Nadu	Miocene	C 56

Fossil No.	Fossil Name	Fossil Group	Formation and area	Formation Age	Formation Number
C 97	Crytorhytis attenuate	Gastropoda	Kulakkalnattam Formation, Trichino-poly Gr, Trichinopoly	Cretaceous	K 69
C 98	Ctensis sp.	Plantae	Intertrappean, Dubrajpur Formation, Rajmahal Basin	Triassic	i 8-9
C 99	Cuma disjuncta	Gastropoda	Chilka Lake deposit of estuarine shells, Orissa	pleistocene	C 33
C 100	Cuneatisporites	PF	Spiti Member, Uttarakhand	Up. Jurassic	S 62
C 101	Cushmanidea bhatiai	Ostracoda	Infrara-trappean & Intertrappean beds, Rajahmundry, Andhra Pradesh	Cretaceous	i 5-1b i 8-1b
C 102	Cyanophyta	Algae	Bagh Beds (Nimar Sandstone) Narmada Basin	Cretaceous	B 8
C 103	Cyathidites minor	PF	Inter-trappean beds of Lalitpur, Uttar Pradesh	Cretaceous	i 8-7a
C 104	Cyathophyllum sp.	Coral	Muth Formation, Spiti	Devonian	M 66
			Lipak Formation	Carboniferous	L 24
C 105	Cyclocrinites	Algae	Takche Formation, Spiti	Silurian-Ordovician	T 3
C 106	Cyclocypris amphiboles	Crustacea	Inter-trappean beds of Kora, Gujarat; Upparhatti, Karnataka; Padwar/Ranipur, Madhya Pradesh	Cretaceous	i 8-2c, 3c, 4c
C 107	Cyclocypris sahnii	Ostracoda	Inter-trappean beds of Kora, Gujarat; Padwar/Ranipur, Madhya Pradesh	Cretaceous	i 8-2c, 4c
C 108	Cyclocypris transitoria		Infra-trappean (=Lameta) beds of Dongargaon and Pisdura, Maharashtra		i 5-4a,b
C 109	Cyclolobus	Cephalopoda	Zewan Formation & Khunamuh Formation, J &K	Permian	Z 4 & K 51
			Productus Shale, Spiti		P 51
C 110	Cyclolobus oldhami		Tethyan Basin, Himalaya		T 20
			Kukti Formation, Lahaul, Himachal Pradesh		K 68

Fossil No.	Fossil Name	Fossil Group	Formation and area	Formation Age	Formation Number
			Gungri Formation, Himachal Pradesh		G 33
			Boulder Slate Formation, Garhwal, Uttarakhand		B 71
C 111	Cypria cyrtonidion	Ostracoda	Inter-trappean beds of Kora, Lakshmipur, Gurmatkal, Mohgaon-Kalan & Nagpur	Cretaceous	i 8-2c,d; 3b; 4b; 5b
C 112	Cypria elata		Infra-trappean (=Lameta) beds of Dongargaon, Maharashtra		i 5-4a
C 113	Cypridea	Ostracoda	Inter-trappean beds of Asifabad, Mumbai, Nagpur, Mamoni, and Gurmatkal	Cretaceous	i 8-1a,5a,5b, 6a,3b
			Infra-trappean (=Lameta) beds of Jabalpur, Dongargaon and Pisdura		i 5-3a, 4a,b
C 114	Cypridopsis astralos	Ostracoda	Inter-trappean beds of Lakshmipur, Gujarat	Cretaceous	i 8-2d
C 115	Cypridopsis bugintsavica ?	Ostracoda	Infra-trappean (=Lameta) beds of Jabalpur, Madhya Pradesh	Cretaceous	i 5-3a
C 116	Cypridopsis elachistos	Ostracoda	Inter-trappean beds of Mohgaon-Kalan, Madhya Pradesh	Cretaceous	i 8-4b
C 117	Cypridopsis hyperectyphos		Inter-trappean beds of Kora and Lakshmipur, Gujarat; Gurmatkal, Karnataka		i 8-2c,d; 3b
C 118	Cypridopsis legitima		Inter-trappean beds of Lakshmipur, Gujarat		i 8-2d
C 119	Cypridopsis sp.		Infra-trappean (=Lameta) beds of Dongargaon, Maharashtra		i 5-4a
C 120	Cypridospis wynnei		Inter-trappean beds of Lakshmipur, Gurmatkal & Nagpur		i 8-2d,3b, 5b
C 121	Cypris cylindrical	Ostracoda	Inter-trappean beds of Bombay, Maharashtra	Cretaceous	I 8-5a
C 122	Cypris semimarginata				

Fossil No.	Fossil Name	Fossil Group	Formation and area	Formation Age	Formation Number
C 123	Cyprois	Ostracoda	Inter-trappean beds, Bombay	Cretaceous	i 8-5a
C124	Cyrtella	Fungi	Agglomeratic Slate, J & K	Carboniferous-Permian	A 4
C 125	Cyrtella nitiensis		Kuling Formation, Uttarakhand	Permian	K 69
C 126	Cytherella mohani	Ostracoda	Infra-trappean beds, Rajahmundry	Cretaceous	i 5-1b
			Inter-trappean beds of Rajahmundry, Andhra Pradesh		i 8-1b
C 127	Cytherella sp.		Khalsi Formation, Jammu and Kashmir		K 42
C 128	Cytherelloidea sp. Cf. C. keiji	Ostracoda	Inter-trappean beds of Rajahmundry, Andhra Pradesh	Cretaceous	i 8-1b
C 129	Cytheridella rajahmundryensis	Ostracode	Infra-trappean beds, Rajahmundry	Cretaceous	i 5-1b
			Inter-trappean beds of Rajahmundry, Andhra Pradesh; Nagpur, Mharashtra; Mamoni, Rajasthan		i 8-1b,5b, 6a
C 130	Cytheridella strangulate		Inter-trappean beds of Asifabad, Andhra Pradesh	Cretaceous	i 8-1a
			Infra-trappean (=Lameta) beds of Dongargaon and Pisdura, Maharashtra	Cretaceous	i 5-4a,b
D 1	Damesia cretacea	Gastropodas	Khalsi Formation, Jammu and Kashmir	Cretaceous	K 42
D 2	Damesites sugota	Ammonite	Anaipadi Formation, Trichinopoly Gr, Trichinopoly	Cretaceous	A 22
D 3	Damesops sheridanorum	Trilobita	Tethyan Basin, Himalaya	Cambrian	T 20
D 4	Damsella	Trilobita	Kunzun La Formation, Zanskar, J&K	Cambrian	K 73b
D 5	Dandotiaspora dilate	PF	Inter-trappean beds of Lalitpur	Cretaceous	i 8-7a
D 6	Dandotiaspora plicata	PF	Inter-trappean beds of Lalitpur	Cretaceous	i 8-7a
D 7	Dandotiaspora pseudoauriculata	PF	Inter-trappean beds of Lalitpur	Cretaceous	i 8-7a

Fossil No.	Fossil Name	Fossil Group	Formation and area	Formation Age	Formation Number
D 8	Daonella indica	Lamellibranchia	Daonella Limestone, Spiti	Triassic	D 8
			Rimkhim Member, Uttarakhand		R 36
D 9	Daonella lammeli		Daonella Shales, Spiti		D 9
D 10	Dapalis bhatiai	Fish	Inter-trappean beds of Rangapur	Cretaceous	i 8-1d
D 11	Darwinula sp.	Ostracoda	Infra-trappean (=Lameta) beds of Jabalpur & Dongargaon,	Cretaceous	I 5-3a,4a
D 12	Darwinula torpedo	Ostracoda	Inter-trappean beds of Gurmatkal		i 8-3b
D 13	Dasielodella savitrae	Algae	Niniyur Formation, Trichinopoly	Paleocene	N 23
D 14	Deccanolestes robustus	Mammalia	Inter-trappean beds of Naskal	Cretaceous	i 8-1c
D 15	Deccanolestes hislopi		Inter-trappean beds of Naskal, Rangapur & Kisalpuri		i 8-1c,d,4a
D 16	Deinotherium	Proboscidea	Siwalik Group, Lower & Middle Divisions	Miocene	S 51a & b
D 17	Delphinus	Mammalia (Dolphin)	Piram Beds, Cambay,Gujarat	Tertiary	P 36
D 18	Deltopecten	Lamellibranchia	Gechang Formation	Permian	G 10
			Umaria Marine Bed and Manendragarh Marine Bed	Permian	U 5, M 22
D 19	Deltopecten mitcheli		Agglomeratic Slate, J & K	Carboniferous-Permian	A 4
			Miri Formation, Arunachal Pradesh	Permian	M 54
D 20	Densiosporites	PF	Umia Formation, Kutch	Cretaceous	U 8
D 21	Densipollenites	PF	Barakar Formation, Pranhita-Godavari Valley	Lr. Permian	B 24
			Gondwana Supergroup, Lower Division	Permian to Triassic	G 23a
D 22	Dentalina sp.	Foraminifera	Chikkim Formation, Himachal Pradesh	Cretaceous	C 31
D 23	Derbyia	Brachiopoda	Agglomeratic Slate, J & K	Carboniferous-Permian	A 4

Fossil No.	Fossil Name	Fossil Group	Formation and area	Formation Age	Formation Number
D 24	Deshayesites bodei	Ammonite	Habur Formation, Jaisalmer, Rajasthan	Cretaceous	H 1
D 25	Diceracardium	Lamellibranchia	Megalodon Limestone, Spiti	Triassic	M 40
D 26	Dichoteris elorensis	Plantae	Tirupati Formation	Lr. Cretaceous	T 34
D 27	Dicoryphochoerus	Mammalia	Tertiary rocks of Kutch, Gujarat	Tertiary	T 18
D 28	Dicroidium sp.	Pteridospermatophyta	Vemavaram Formation	Cretaceous	V 7
D 29	Dictyoconoides	Foraminifera	Dras Volcanics, Jammu and Kashmir	Cretaceous	D 49
D 30	Dictyophyllum saporta	Fern	Infra-trappean beds, Jabalpur	Cretaceous-Paleocene	i 5-3a
D 31	Dictyozamites indicus	Gymnospermae	Rajmahal Traps	Cretaceous	R 9
			Tirupati Formation		T 34
D 32	Dictyozamites sp.		Vemavaram Formation		V 7
D 33	Didymaulichnus	TF	Kunzun La Formation, Spiti	Lr. Cambrian	K 73a
D 34	Dielasma	Brachiopoda	Po Formation, Spiti-Kinnaur, Himachal Pradesh	Carboniferous	P 38
			Badhaura Formation, Bikaner-Nagaur Basin, Rajasthan	Permian	B 5
			Talchir Formation, Daltonganj sub-basin, Damodar Basin	Permian-Carboniferous	T 6
D 35	Dielasma dadanese		Miri Formation, Arunachal Pradesh	Permian	M 54
D 36	Dielasma julicum		Tropites Beds, Spiti	Triassic	T 47
D 37	Dielasma kinauricum		Boulder Slate Formation, Garhwal, Uttarakhand	Permian	B 71
D 38	Dielasma latouche		Boulder Slate Formation, Garhwal, Uttarakhand	Permian	B 71
D 39	Dimorphichnus	TF	Kunzun La Formation, Spiti	Lr. Cambrian	K 73a
D 40	Dimorphichnus	TF	Kunzun La Formation, Spiti	Lr. Cambrian	K 73a
D 41	Diplagnostus planicauda	Trilobita	Tethyan Basin, Himalaya	Cambrian	T 20

Fossil No.	Fossil Name	Fossil Group	Formation and area	Formation Age	Formation Number
D 42	Diplichnites	TF	Kunzun La Formation, Spiti	Lr. Cambrian	K 73a
D 43	Dipterocarpceae	Plantae (Angiospermae)	Cuddalore Sandstone Formation, Tamil Nadu	Miocene	C 56
D 44	Discocyclina sps.	Foraminifera	Dinod Formation, Gujarat	Eocene	D 40
D 45	Discoglossidae indet	Amphibian	Inter-trappean beds of Naskal, Andhra Pradesh	Cretaceous	i 8-1c
D 46	Discorbis	Foraminifera	Khari Series, Kutch	Miocene	K 44
D 47	Discosphinctes	Ammonite	Dhosa Oolite, Chari Formation, Kutch	Jurassic	D 34
D 48	Distoloceras	Ammonite	Spiti Shale Formation, Malla Johar Supergroup	Lr. Cretaceous	S 61
D 49	Docorypho-choerus	Suidae	Siwalik Group, Upper Division	Pleistocene	S 51c
D 50	Dogaddanella audeni	Bryozoa	Boulder Slate Formation, Garhwal, Uttarakhand	Permian	B 71
D 51	Dolichocephalites	Cephalopoda	Macrocephalus beds, Chari Formation, Kutch	Jurassic	M 2
			Lapthal Member, Uttarakhand	Triassic	L 13
D 52	Dorcadoxa	Bovidae	Siwalik Group, Lower Division	Miocene	S 51a
D 53	Dorsoplanites	Cephalopoda	Upper Katrol Shales, Kutch	Jurassic	U 15
D 54	Fossil deleted due to lack of conformity				
D 55	Durania mutabilis	Lamellibrachia	Ottakovil Formation, Ariyalur Gr, Trichinopoly	Cretaceous	O 9
E 1	Echitricolporites maristellae	PF	Inter-trappean beds of Padwar/Ranipur, Madhya Pradesh	Cretaceous	i 8-4c
E 2	Elaphas hysudricus	Proboscidea	Pleistocene sediments of Central Narmada (upper group), Palmar	Pleistocene	P 37b
E 3	Elathia	Trilobita	Kunzun La Formation, Zanskar, J&K	Cambrian	K 73b
E 4	Elatocladus	Coniferales	Gangapur Formation, Umia Beds, Tirupati Formation	Cretaceous	G 6, U 7, T34
E 5	Elatocladus conferta		Terani Plant Beds	Lr. Cretaceous	T 14
			Pariwar Formation, Jaisalmer Rajasthan	Cretaceous	P 23

Fossil No.	Fossil Name	Fossil Group	Formation and area	Formation Age	Formation Number
E 6	Elatocladus plana		Sivaganga Beds, Tamil Nadu	Lr. Cretaceous	S 49
E 7	Elephas antiques (bones)	Proboscidea	Quaternary sediments of Maharashtra, Godavari valley between Nasik and Paithan	Quaternary	Q 8
E 8	Elephas hysudncus	Proboscidea	Pleistocene sediments of Kurnool, Ghod, Manjra, Pravara and Godavari	Pleistocene	P 37a
E 9	Elephas hysudricus	Proboscidea	Pleistocene sediments of Central Narmada (upper group), Palmar	Pleistocene	P 37b
			Pleistocene sediments of Lower Karewa, Pinjor of Upper Siwaliks	Pleistocene	P 37d
			Quaternary sediments of central India	Lr. to Up. Pleistocene	Q 4
			Quaternary sediments of Tamil Nadu, Soyamali	Pleistocene	Q 9
E 10	Elephas maximus		Pleistocene sediments of Kurnool, Ghod, Manjra, Pravara and Godavari	Pleistocene	P 37a
			Quaternary sediments of central India	Up. Pleistoce-ne to Holocene	Q 4
E 11	Elephas namadicus		Quaternary sediments of central India	Mid. Pleistocene	Q 4
E 12	Elphidium	Foraminifera	Broach Formation, Gujarat	Pliocene	B 72
			Quaternary sediments of Gujarat, Mainland	Quaternary	Q 5
			Miliolite / Porbandar Stone, Gujarat	Pleistocene	M 49
E 13	Emekops	Trilobita	Kunzun La Formation, Zanskar, J&K	Cambrian	K 73b
E 14	Eomiodon	Lamellibranchia	Pachcham basal beds, Kutch	Jurassic	P 1
E 15	Eoschizodus	Lamellibranchia	Yamne Formation, Arunachal Pradesh	Permian	Y 2
E 16	Etheripecten				
E 17	Epimayaites	Ammonite	Kanthkot Sandstone, Kutch	Jurassic	K 23

Fossil No.	Fossil Name	Fossil Group	Formation and area	Formation Age	Formation Number
			Dhosa Oolite, Kutch		D 34
			Spiti Formation, Himachal Pradesh	Up. Jurassic to Lr. Cretaceous	S 61
E 18	Episageceras dalailamiae	Cephalopoda	Rambakot Member		R 13
E 19	Equiseloto-sporites	Plantae	Inter-trappean beds of Asifabad, Nagpur, Mamoni and Gurmatkal	Cretaceous	i 8-1a,5b, 6a,3b
E 20	Equisetites	Pteridophyta	Umia Beds, Kutch, Gujarat	Cretaceous	U 7
E 21	Equus asimus	Equidae	Quaternary sediments of central India	Up. Pleistocene to Holocene	Q 4
E 22	Equus namadicus	Equidae	Pleistocene sediments of Central Narmada (upper group), Palmar	Pleistocene	P 37b
E 23	Equus namadicus	Equidae	Quaternary sediments of central India	Mid. to Up. Pleistocene	Q 4
E 24	Equus namadicus (tooth)	Equidae	Quaternary sediments of Maharashtra, Godavari valley between Nasik and Paithan	Quaternary	Q 8
E 25	Equus sivalensis	Equidae	Pleistocene sediments of Lower Karewa & Upper Siwalik Group, Pinjor	Pleistocene	P 37d
E 26	Equus sps. (body parts)	Equidae	Quaternary sediments of Andhra Pradesh, Gunglakamma and Munneru river basins	Pleistocene	Q 3
E 27	Ethelia alba	Algae	Kallakkudi Limestone Formation, Uttatur Gr, Trichinopoly	Cretaceous	K 11
E 28	Eubaculites	Ammonite	Ariyalur Group, Trichinopoly	Cretfaceous	A 35
E 29	Eubaculites otacodensis		Anaipadi Formation, Trichinopoly Gr, Trichinopoly		A 22
E 30	Eucalycoceras pentagonum	Ammonite	Karai Shale Formation and Maravattur Clay Formation	Cretaceous	K 26 M 28
E 31	Eucandona kakamorpha	Ostracoda	Inter-trappean beds of Gurmatkal, Karnataka	Cretaceous	i 8-3b

Fossil No.	Fossil Name	Fossil Group	Formation and area	Formation Age	Formation Number
E 32	Eucypris cf. E. bajshintsavica	Ostracoda	Infra-trappean (=Lameta) beds of Jabalpur, Madhya Pradesh	Cretaceous	i 5-3a
E 33	Eucypris intervolcanus		Inter-trappean beds of Kora and Lakshmipur, Gujarat; Gurmatkal, Karnataka; Mohgaon-Kalan, Madhya Pradesh		i 8-2c,d; 3b; 4b
E 34	Eugyridae	Coral	Kallakkudi Limestone Formation, Uttatur Gr, Trichinopoly	Cretaceous	K 11
E 35	Euomphalus	Gastropoda	Umaria Marine Bed and Manendragarh Marine Bed	Permian	U 5; M 22
E 36	Eurydesma	Lamellibranchia	Umaria Marine Bed, Manendragarh Marine Bed & Talchir Formation	Permian	U 5, M 22 & T 6
E 37	Eurydesma cordatum		Tethyan Basin, Himalaya	Permian	T 20
			Agglomeratic Slate, J & K	Carboniferous-Permian	A 4
			Miri Formation, Arunachal Pradesh	Permian	M 54
			Gechang Formation	Permian	G 10
E 38	Eurydesma globosum		Talchir Formation, Daltonganj sub-basin, Damodar Basin	Permian-Carboniferous	T 6
E 39	Eurydesma hasdoensis		Gechang Formation	Permian	G 10
E 40	Eurydesma hovbartene		Talchir Formation, Daltonganj sub-basin, Damodar Basin	Permian-Carboniferous	T 6
E 41	Eurydesma manindragarhensis		Gechang Formation	Permian	G 10
E 42	Eurydesma playfordi		Talchir Formation, Daltonganj sub-basin, Damodar Basin	Permian-Carboniferous	T 6
E 43	Exogyra tamulica	Lamellibranchia	Ariyalur Gr. Trichinopoly	Cretaceous	A 35
			Kallankurichi Formation, Trichinopoly	Cretaceous	K 13

Fossil No.	Fossil Name	Fossil Group	Formation and area	Formation Age	Formation Number
F 1	Faunipollenites	PF	Gondwana Supergroup, Lower Division	Permian to Triassic	G 23a
			Barakar Formation, Pranhita-Godavari Valley	Lr. Permian	B 24
F 2	Favosites	Coral	Takche Formation, Spiti	Silurian-Ordovician	T 3
			Variegated Formation, Uttarakhand	Silurian-Ordovician	V 5
			Muth Formation, Byans, Uttarakhand	Devonian	M 66
F 3	Felis	Carnivora	Siwalik Group, Upper Division	Pleistocene	S 51c
F 4	Fenestella	Bryozoa	Takche Formation, Spiti	Silurian-Ordovician	T 3
			Fenestella Shale Formation, Kashmir	Carboniferous	F 2
			Saser Brangsa Formation	Carboniferous	S 13
			Po Formation, Spiti-Kinnaur area	Carboniferous	P 38
			Miri Formation, Arunachal Pradesh	Permian	M 54
F 5	Fenestella garhwalensis		Boulder Slate Formation	Permian	B 71
F 6	Fenestella sp. Aff. Jabbiensis		Talchir Formation, Daltonganj sub-basin, Damodar Basin	Permian-Carboniferous	T 6
F 7	Flemingites rohilla	Cephalopoda	Hedenstroemia Beds, Spiti	Triassic	H 12
			Rambakot Member	Triassic	R 13
F 8	Frambocythere tumiensis anjarensis	Ostracoda	Inter-trappean beds of Anjar, Gujarat; Gurmatkal, Karnataka	Cretaceous	i 8-2a;3b
			Infra-trappean (=Lameta) beds of Jabalpur, Madhya Pradesh		i 5-3a
F 9	Furnishiana	Conodonta	Kunzun La Formation, Spiti	Cambrian	K 73a
F 10	Fusispirites plicatus	NA	Agglomeratic Slate, J & K	Carboniferous-Permian	A 4

Fossil No.	Fossil Name	Fossil Group	Formation and area	Formation Age	Formation Number
G 1	Gabonisporis bacaricumulus	PF	Inter-trappean beds of Anjar, Gujarat	Cretaceous	i 8-2a
G 2	Gabonisporis vigourouxii		Inter-trappean beds of Asifabad, Nagpur, Mamoni and Gurmatkal		i 8-1a;5b;6a;3b
G 3	Gabonisporis vigourouxii		Inter-trappean beds of Naskal, Andhra Pradesh; Anjar, Gujarat; Mohgaon-Kalan and Padwar/ Ranipur, Madhya Pradesh		i 8-1c;2a;4b,c
G 4	Gaindatherium	Rhinocerotidae	Siwalik Group, Middle Division	Pliocene	S 51b
G 5	Gangamopteris	Pteridospermatophyta	Barakar Formation, Pranhita-Godavari Valley	Lr. Permian	B 24
G 6	Gangamopteris		Gondwana, Extra-Peninsular India, Arunachal Pradesh	Permian	G 22
G 7	Gangamopteris angustifolia		Gangamopteris Beds, Gondwana, Kashmir	Permian	G 5
G 8	Gangamopteris angustifolia		Gondwana Supergroup, Lr. Div.	Permian to Triassic	G 23a
G 9	Gangamopteris angustifolia		Nishatbagh Bed, J & K	Lr. Permian	N 25
G 10	Gangamopteris cf. cyclopteroides		Bahreli Formation, Arunachal Pradesh	Permian	B 10
G 11	Gangamopteris cyclopteroides		Gondwana Supergroup, Lower Division	Permian to Triassic	G 23a
G 12	Gangamopteris equioelean		Chhongtash Formation, J&K	Permian	C 27
G 13	Gangamopteris kashmirensis		Gangamopteris Beds, Kashmir	Permian	G 5
G 14	Gangamopteris kashmirensis		Nishatbagh Bed, J & K	Lr. Permian	N 25
G 15	Gazella	Bovidae	Siwalik Group, Middle Division	Pliocene	S 51b
G 16	Gigantopithecus	Primates	Siwalik Group, Upper Division	Pleistocene	S 51c
G 17	Ginglymostoma sokotense	Fish (Shark teeth)	Bagh Beds (Amba Dongar) Narmada Basin	Cretaceous	B 8
G 18	Ginkgoites crassipes	Gymnospermae	Sivaganga Beds, Tamil Nadu	Lr. Cretaceous	S 49

Fossil No.	Fossil Name	Fossil Group	Formation and area	Formation Age	Formation Number
G 19	Ginkgophyllum haydeni	Ginkogales	Mamal Formation, J & K	Lr. Permian	M 19
G 20	Giraffokeryx	Giraffidae	Siwalik Group, Middle Division	Pliocene	S 51b
G 21	Girvanella	Algae	Takche Formation, Spiti	Silurian-Ordovician	T 3
G 22	Gleichenia	Fern	Gangapur Formation.	Lr. Cretaceous	G 6
			Umia Beds, Kutch, Gujarat	Cretaceous	U 7
G 23	Gleichenites gleichenoides	Fern	Dubrajpur Formation	Triassic	D 51
			Tirupati Formation	Lr. Cretaceous	T 34
G 24	Gleichniidites	Fern	Chari Formation and Katrol Formation, Kutch	Jurassic	C 16 & K 38
G 25	Gleichniidites cicatricosisparites		Umia Stage, Kutch	Cretaceous	U 11
G 26	Globanomalina wilocoxensis	Foraminifera	Kakdi Stage, Kutch	Eocene	K 5
G 27	Globigerina aqueinsis	Foraminifera	Kakdi Stage, Kutch	Eocene	K 5
G 28	Globigerina bulloides		Miliolite / Porbandar Stone, Gujarat	Pleistocene	M 49
G 29	Globigerina trilobata		Miliolite / Porbandar Stone, Gujarat	Pleistocene	M 49
G 30	Globigerinelloides asperus	Foraminifera	Ariyalur Gr. Trichinopoly	Cretaceous	A 35
			Sillakudi Formation	Cretaceous	S 37
G 31	Globigerinoides	Foraminifera	Qazil Langer Formation, J & K	Cretaceous	Q 1
G 32	Globorotalia	Foraminifera	Niniyur Formation, Trichinopoly	Paleocne	N 23
G 33	Globorotalia cf. G. menardii	Foraminifera	Inter-trappean beds of Rajahmundry, Andhra Pradesh	Cretaceous	i 8-1b
G 34	Globorotalia imitate		Kakdi Stage, Kutch	Eocene	K 5
G 35	Globorotalia lehneri		Babia Stage, Kutch	Eocene	B 3
G 36	Globorotalia trobulosa		Kakdi Stage, Kutch	Eocene	K 5
G 37	Globotruncana	Foraminifera	Dras Volcanics, Jammu and Kashmir	Cretaceous	D 49

Fossil No.	Fossil Name	Fossil Group	Formation and area	Formation Age	Formation Number
			Mahdek Formation, Assam/Meghalaya	Cretaceous	M 4
G 38	Globotruncana gansseri dicarinata		Chikkim Formation	Cretaceous	C 31
G 39	Globotruncana arca		Ottakovil Formation, Ariyalur Gr.	Cretaceous	O 9
			Infra-trappean beds, Rajahmundry	Cretaceous	i 5-1b
G 40	Globotruncana concavata		Baratang Group, Andaman and Nicobar Islands	Cretaceous	B 25
G 41	Globotruncana gansseri		Kallankurichi Formation	Cretaceous	K 13
G 42	Globotruncana linneiana		Chikkim Formation	Cretaceous	C 31
G 43	Globotruncana stauriformis		Ottakovil Formation, Ariyalur Gr,	Cretaceous	O 9
G 44	Globotruncana stuarti		Infra-trappean beds, Rajahmundry	Cretaceous	i 5-1b
G 45	Globotruncana tricarinata		Chikkim Formation	Cretaceous	C 31
G 46	Globotruncanella citae	Foraminifera	Infra-trappean beds, Rajahmundry	Cretaceous	i 5-1b
G 47	Glomospira	Foraminifera	Umaria Marine Bed and Manendragarh Marine Bed	Permian	U 5; M 22
G 48	Glossopteris sp	Pteridospermatophyta	Talchir Formation and Barakar Formation, Gondwana SG	Permian	T 6
			Bijori Formation, Gondwana SG	Permian	B 58
			Kota Formation	Jurassic	K 62
			Gondwana, Extra-Peninsular India, Arunachal Pradesh	Permian	G 22
			Bahreli Formation, Arunachal Pradesh	Permian	B 10
			Barakar Formation, Pranhita-Godavari Valley	Lr. Permian	B 24
G 49	Glossopteris angustifolia		Gangamopteris Beds, Kashmir	Permian	G 5
			Raniganj Formation	Permian	R 23

Fossil No.	Fossil Name	Fossil Group	Formation and area	Formation Age	Formation Number
G 50	Glossopteris browniana var. indica		Raniganj Formation	Permian	R 23
G 51	Glossopteris communis		Gangamopteris Beds, Kashmir	Permian	G 5
G 52	Glossopteris feistmantelli		Mamal Formation, J & K	Lr. Permian	M 19
G 53	Glossopteris indica		Gangamopteris Beds, Kashmir	Permian	G 5
			Bahreli Formation, Arunachal Pradesh	Permian	B 10
			Gondwana Supergroup, Lower Division	Permian to Triassic	G 23a
G 54	Glossopteris intermittens		Mamal Formation, J & K	Lr. Permian	M 19
G 55	Glossopteris longicaulis		Bahreli Formation, Arunachal Pradesh	Permian	B 10
			Nishatbagh Bed, J & K	Lr. Permian	N 25
G 56	Glossopteris nishatbaghensis		Nishatbagh Bed, J & K	Lr. Permian	N 25
G 57	Glossopteris sahni		Mamal Formation, J & K	Lr. Permian	M 19
G 58	Gomphocythere akalypton	Ostracoda	Inter-trappean beds of Gurmatkal, Karnataka	Cretaceous	i 8-3b
G 59	Gomphocythere dasyderma				
G 60	Gomphocythere paucisulcatus		Inter-trappean beds of Mohgaon-Kalan, Madhya Pradesh		i 8-4b
G 61	Gomphocythere strangulate		Inter-trappean beds of Gurmatkal, Karnataka		i 8-3b
G 62	Gomphotherium	Proboscidea	Siwalik Group, Lower Division	Miocene	S 51a
G 63	Gondwanablatta reticulata	Insecta	Gangamopteris Beds	Permian	G 5
G 64	Gondwanosaurus bijoriensis	Labyrint-hodontia	Bijori Formation, Gondwana SG	Permian	B 58
G 65	Goniopora	Lamelli-branchia	Yamne Formation, Arunachal Pradesh	Permian	Y 2
G 66	Grambastichara sp.	Algae	Inter-trappean beds of Gurmatkal, Karnataka	Cretaceous	I 8-3b

Fossil No.	Fossil Name	Fossil Group	Formation and area	Formation Age	Formation Number
G 67	Grayiceras koeneni	Cephalopoda	Spiti Formation, Spiti	Up. Jurassic to Lr. Cretaceous	S 61
G 68	Gryphea sp	Lamellibranchia	Ariyalur Group, Trichinopoly	Cretaceous	A 35
G 69	Gryphea aff. Baylei	Lamellibranchia	Giumal Formation, Himachal Pradesh	Cretaceous	G 15
G 70	Guembelina globifera	Foraminifera	Inter-trappean beds of Rajahmundry, Andhra Pradesh	Cretaceous	i 8-1b
G 71	Guembelitria colombiana	Foraminifera	Babia Stage, Kutch	Eocene	B 3
G 72	Gumbelina	Foraminifera	Mahadek Formation, Assam/Meghalaya	Cretaceous	M 4
G 73	Gunnarites kaklika	Ammonite	Ottakovil Formation, Ariyalur Gr.	Cretaceous	O 9
G 74	Guttulina lacteal	Foraminifera	Infra-trappean (=Lameta) beds of Rajahmundry, Andhra Pradesh	Cretaceous	i 5-1b
G 75	Gymnodiscoceras griesbachi	Ammonite	Spiti Formation	Jurassic	S 61
G 76	Gymnodiscoceras lymani	Ammonite	Spiti Formation	Jurassic	S 61
G 77	Gypsins globules	Foraminifera	Khari Series, Kutch	Miocene	K 44
G 78	Gypsins globules	Foraminifera	Babia Stage, Kutch	Eocene	B 3
G 79	Gyrochorte	TF	Garbyang Formation	Cambrian	G 8
G 80	Gyrochorte	TF	Po Formation, Spiti-Kinnaur	Carboniferous	P 38
G 81	Gyrodes tenellus	Gastropoda	Trichinopoly Group	Cretaceous	T 44
G 82	Gyrodestenellus	Gastropoda	Kulakkalnattam Formation	Cretaceous	K 69
G 83	Gyrogonites medicaginula	Algae	Inter-trappean beds of Biligi and Gurmatkal, Karnataka	Cretaceous	i 8-3a,b
H 1	Halimeda	Algae	Uttatur Gr. Trichinopoly	Cretaceous	U 21
H 2	Hallopora	Bryozoa	Takche Formation, Spiti	Silurian-Ordovician	T 3
H 3	Halobia cf. comata	Lamellibranchia	Halobia Beds, Spiti Rimkhim Member, Uttarakhand	Triassic	H 4 R 36
H 4	Halorites procyon	Cephalopoda	Kioto Member, Uttarakhand	Triassic	K 55

Fossil No.	Fossil Name	Fossil Group	Formation and area	Formation Age	Formation Number
H 5	Halysites	Coral	Takche Formation, Spiti	Ordovician-Silurian	T 3
H 6	Halysities catenularia var. kanurensis	Coral	Tethyan Basin, Himalaya	Silurian	T 20
H 7	Haploceras	Cephalopoda	Gajansar Beds, Kutch	Jurassic	G 2
H 8	Haplophragmoides	Foraminifera	Raghvapuram Formation	Lr. Cretaceous	R 2
			Sriperumbudur Formation	Mid. Jurassic (?)	S 63
			Surma Group, NE India	Early Miocene	S 73
H 9	Harrisichara muricata	Charophyta (Algae)	Inter-trappean beds of Kora, Gujarat; Gurmatkal, Karnataka	Cretaceous	i 8-2c,3b
H 10	Harungela sahnii	Fish	Inter-trappean beds of Rangapur, Andhra Pradesh	Cretaceous	i 8-1d
H 11	Hauriceras rambda	Ammonite	Kallankurichi Formation, Ariyalur Gr, Trichinopoly	Cretaceous	K 13
H 12	Hedbergella planispira	Foraminifera	Kallakkudi Limestone Formation, Uttatur Formation, Trichinopoly	Cretaceous	K 11
H 13	Hedbergella sp.	Foraminifera	Chikkim Formation, Himachal Pradesh	Cretaceous	C 31
H 14	Hedemstroemia mojsisovicsi	Cephalopoda	Hedenstroemia Beds, Spiti	Triassic	H 12
H 15	Heliolites	Coral	Takche Formation, Spiti	Silurian-Ordovician	T 3
H 16	Helmenthopsis	TF	Tethyan Basin, Himalaya	Late Precambrian	T 20
H 17	Helvetoglobo-truncana Helvetica	Foraminifera	Karai Shale Formation and Maravattur Clay Formation	Cretaceous	K 26 M 28
H 18	Hemiaster	Echinoidea	Bagh Beds, Narmada Basin	Cretaceous	B 8
H 19	Hemicorpus	Plantae	Inter-trappean beds of Asifabad, Nagpur, Mamoni and Gurmatkal	Cretaceous	i 8-1a,5b,6a&3b
H 20	Hemilytoceras	Ammonite	Umia ammonite bed, Kutch	Cretaceous	U6

Fossil No.	Fossil Name	Fossil Group	Formation and area	Formation Age	Formation Number
H 21	Hemimeryx	Anthracoth-eriidae	Siwalik Group, Lower Division	Miocene	S 51a
H 22	Hermanites sastryi	Ostracoda	Inter-trappean beds of Rajahmundry, Andhra Pradesh	Cretaceous	i 8-1b
H 23	Hermanites sp.	Ostracoda	Inter-trappean beds of Rajahmundry, Andhra Pradesh; Kora, Gujarat	Cretaceous	i 8-1b,2c
H 24	Heterohelix	Foraminifera	Dras Volcanics, Jammu and Kashmir	Cretaceous	D 49
H 25	Heteroitis	NA	Takli Formation, Nagpur	Cretaceous-Paleocene	T 4
H 26	Heterostegina sp.	Foraminifera	Ramania Stage, Kutch	Oligocene	R 12
H 27	Hexaprotodon	Artiodactyla	Pleistocene sediments of Kurnool, Ghod, Manjra, Pravara and Godavari	Pleistocene	P 37a
H 28	Hexaprotodon	Artiodactyla	Pleistocene sediments of Central Narmada (upper group), Palmar	Pleistocene	P 37-b
H 29	Hexaprotodon namadicus	Artiodactyla	Quaternary sediments of central India	Mid. Pleistocene	Q 4
H 30	Hexaprotodon palaeindicus	Artiodactyla	Quaternary sediments of central India	Up. pleistocene	Q 4
H 31	Hildoglochiceras	Ammonite	Gajansar Beds, Kutch	Jurassic	G 2
H 32	Hildoglochiceras kobelli		Spiti Formation	Jurassic	S 61
H 33	Hildoglochiceras virgatisphinctes		Spiti Member, Uttarakhand	jurassic	S 62
H 34	Hildoqlochiceras		Upper Katrol Shales, Kutch	Jurassic	U 15
H 35	Himalayites	Ammonite	Spiti Member, Uttarakhand	jurassic	S 62
H 36	Himalayites gr. seideli	Ammonite	Spiti Formation	Jurassic	S 61
H 37	Hindeodus cristulus?	Conodonta	Muth Formation, Spiti	Devonian	M 66
H 38	Hindeodella subtilis	Conodonta	Syringothyris Lst. Formation, Kashmir	Carboniferous	S 79
H 39	Hipparion antelopinum	Equidae	Siwalik Group, Middle Division	Pliocene	S 51b
H 40	Hippohyus	Suidae	Siwalik Group, Upper Division	Pleistocene	S 51c

Fossil No.	Fossil Name	Fossil Group	Formation and area	Formation Age	Formation Number
H 41	Hippopotamus	Anthracot-heriidae	Siwalik Group, Upper Division	Pleistocene	S 51c
H 42	Hippopotamus (Tetraprotodon) palaeindicus (lower jaw)	Artiodactyla	Quaternary sediments of Maharashtra, Godavari valley between Nasik and Paithan	Quaternary	Q 8
H 43	Hippopotamus namadicus	Artiodactyla	Conglomerate Bed, Narmada valley	Middle Pleistocene	C 46
H 44	Hippurites	Lamellibranchia	Dras Formation, Jammu and Kashmir	Cretaceous	D 48
H 45	Holcophylloceras aff. mesolcum	Ammonite	Spiti Formation	Jurassic	S 61
H 46	Holimeda	Algae	Kallakkudi Limestone Formation, Trichinopoly	Cretaceous	K 11
H 47	Homo erectus	Primates (Homo-man)	Conglomerate Bed, Narmada valley	Middle Pleistocene	C 46
H 48	Horaclupea intertrappus	Fish	Inter-trappean beds of Bamanbor, Gujarat	Cretaceous	i 8-2b
H 49	Hornibrookella tewarii	Ostracoda	Inter-trappean beds of Rajahmundry, Andhra Pradesh	Cretaceous	i 8-1b
H 50	Hubertoceras	Cephalopoda	Anceps beds, Kutch	Jurassic	A23
H 51	Hundwavella	Trilobita	Kunzun La Formation, Zanskar, J&K	Cambrian	K 73b
H 52	Hybonoticeras hybonotum	Ammonite	Spiti Formation	Jurassic	S 61
H 53	Hydaspitherium	Giraffidae	Siwalik Group, Middle Division	Pliocene	S 51b
H 54	Hydnocarpoxylon bande & khatri	Plantae	Infra-trappean beds, Jabalpur	Cretaceous-Paleocene	i 5-3a
H 55	Hynoticeras	NA	Lower Katrol, Kutch	Jurassic	L 29
H 56	Hyolithes (Orthotheca) aff. Plicatus	Gastropoda	Tethyan Basin, Himalaya	Cambrian	T 20
H 57	Hyperammina	Foraminifera	Umaria Marine Bed and Manendragarh Marine Bed	Permian	U 5 & M 22
H 58	Hystrichos-phaeridium	Plantae (Microflora)	Umia Stage (Ghuneri Member), Kutch, Gujarat	Cretaceous	U 11

Fossil No.	Fossil Name	Fossil Group	Formation and area	Formation Age	Formation Number
H 59	Hystrichosphaeris	Plantae (Microflora)	Umia Stage (Ghuneri Member), Kutch, Gujarat	Cretaceous	U 11
H 60	Hystrix	Rodentia	Siwalik Group, Upper Division	Pleistocene	S 51c
i 1	Ichnogenus Gordia sp. Cf. G. marina	TF	Krol Gr., Nainital Syncline	Neoproterozoic	K 64
i 2	Icriodus arconensis	Conodonta	Tethyan Basin, Himalaya	Devonian	T 20
i 3	Icriodus difficilis				
i 4	Idiostroma	Porifera	Muth Formation, Spiti	Devonian	M 66
i 5	Igdabatis indicus	Fish	Inter-trappean beds of Naskal, Andhra Pradesh; Kisalpuri, Madhya Pradesh, Nagpur, Maharashtra	Cretaceous	i 8-1c;4a;5b
i 6	Impardecispora	PF	Umia Formation, Kutch	Cretaceous	U 8
			Raghvapuram Formation		R 2
i 7	Inapertusporites kedvesii	PF	Inter-trappean beds of Lalitpur, Uttar Pradesh	Cretaceous	I 8-7a
i 8	Inceramus	Lamellibranchia	Sillakudi Formation, Trichinopoly	Cretaceous	S 37
			Sivaganga Formation, Tamil Nadu		S 50
i 9	Idiocycloceras	Cephalopoda	Rehmanni beds, Kutch	Jurassic	R 31
i 10	Indiostroma	Fungi	Lipak Formation	Carboniferous	L 24
i 11	Indobatrachus pusillus	Amphibian	Inter-trappean beds of Bombay, Maharashtra	Cretaceous	i 8-5a
i 12	Indocephalites	Cephalopoda	Macrocephalus beds, Kutch	Jurassic	M 2
i 13	Indophis sahnii	Reptilia	Inter-trappean beds of Naskal, Andhra Pradesh; Anjar, Gujarat	Cretaceous	i 8-1c;2a
i 14	Indosphinctes	Cephalopoda	Anceps beds, Kutch	Jurassic	A 23
i 15	Indotriradiites	Plantae	Barakar Formation, Pranhita-Godavari Valley	Lr. Permian	B 24
i 16	Inoceramus	Lamellibranchia	Ariyalur Gr. Trichinopoly	Cretaceous	A 35
			Chikkim Formation		C 31
			Bagh Beds, Narmada Basin		B 8
			Vemavaram Formation		V 7

Fossil No.	Fossil Name	Fossil Group	Formation and area	Formation Age	Formation Number
			Coastal Gondwana Formations, Ramnathpuram, Tamil Nadu		C 45
i 17	Inoceramus cf. sularum		Spiti Formation, Spiti	Up. Jurassic to Lr. Cretaceous	S 61
i 18	Isopodichnus	TF	Garbyang Formation, Uttarakhand	Cambrian	G 8
i 19	Istherium subathuensis	Vertebrata	Subathu Formation, Himachal Pradesh	Paleogene	S 65
J 1	Japonites sugriva	Cephalopoda	Gushal Formation, Himachal Pradesh	Triassic	G 35
J 2	Joannites cymbiformis	Cephalopoda	Grey Beds, Spiti	Triassic	G 28
J 3	Joannites thanamensis		Halobia Beds, Spiti		H 4
J 4	Juvavites angulatus	Cephalopoda	Juvavites Beds, Spiti	Triassic	J 22
K 1	Kanisamys	Rodentia	Siwalik Group, Middle Division	Pliocene	S 51b
K 2	Kamptocephalites	Cephalopoda	Macrocephalus beds, Kutch	Jurassic	M 2
K 3	Karapadites karapadensis	Ammonite	Ariyalur Gr. Trichinopoly	Cretaceous	A 35
K 4	Karapadites karapadensis	Ammonite	Sillakudi Formation	Cretaceous	S 37
K 5	Kashmiroblatta marhomensis	Insecta	Gangamopteris Beds	Permian	G 5
K 6	Kashmiropteris meyenii	Pteridophyta	Gangamopteris Beds	Permian	G 5
K 7	Kathina minuscula	Foraminifera	Sillakudi Formation	Cretaceous	S 37
K 8	Katroliceras	Ammonite	Middle Katrol, Kutch	Jurassic	M 45
K 9	Katroliceras deprecssune		Bhadasar Formation and Baisakhi Formation, Jaisalmer		B 42& B 14
K 10	Katroliceras pottingeri		Spiti Formation		S 61
K 11	Kawizophyllum dunpathriensis	Cordaitales	Gangamopteris Beds	Permian	G 5
K 12	Kellawaysites	Cephalopoda	Rehmanni beds, Kutch	Jurassic	R 31
K 13	Keyserlingites dieneri	Cephalopoda	Hoti Member, Uttarakhand		H 20

Fossil No.	Fossil Name	Fossil Group	Formation and area	Formation Age	Formation Number
			Lower Muscelkalk, Spiti	Triassic	L 31
K 14	Kilianella sp.	Ammonite	Giumal Formation	Cretaceous	G 15
K 15	Kinkeliniceras	Cephalopoda	Anceps beds, Kutch	Jurassic	A 23
K 16	Koninckites cf. yudihisthiara	Cephalopoda	Gushal Formation	Triassic	G 35
K 17	Kossmatia aff. Koeneni	Ammonite	Spiti Formation	Jurassic	S 61
K 18	Kossmatia desmidoptycha		Spiti Formation	Jurassic	S 61
K 19	Kossmatia tenuistriata		Spiti Formation	Jurassic	S 61
K 20	Kosmaticeras recurrens	Ammonite	Kulakkalnattam Formation	Cretaceous	K 69
K 21	Kossmaticeras theoboldinum		Anaipadi Formation		A 22
K 22	Kotasaurus yamanapalliensis	Dinosauria	Kota Formation	Jurassic	K 62
K 23	Krishnapithecus	Primates	Siwalik Group, Middle Division	Pliocene	S 51b
L 1	Lacertilian indet	Reptilia	Inter-trappean beds of Naskal, Andhra Pradesh; Kisalpuri, Madhya Pradesh and Nagpur, Maharashtra	Cretaceous	i 8-1c; 4a; 5b
L 2	Lactariidarum ellipticus	Fish	Inter-trappean beds of Rangapur, Andhra Pradesh	Cretaceous	i 8- 1d
L 3	Lagena sp.	Foraminifera	Chikkim Formation, Himachal Pradesh	Cretaceous	C 31
L 4	Lakiapollis ovatus	PF	Inter-trappean beds of Lalitpur, Uttar Pradesh	Cretaceous	i 8- 7a
L 5	Lamma mongraensis	Fish (Shark teeth)	Bagh Beds (Amba Dongar) Narmada Basin	Cretaceous	B 8
L 6	Lamnimargus himalayaensis	Brachiopoda	Tethyan Basin, Himalaya	Permian	T 20
L 7	Laptaena	Brachiopoda	Ordovician -Silurian rocks of Shans Abari, Kashmir	Ordovician	O 7
L 8	Laptelloidea	Brachiopoda	Ordovician -Silurian rocks of Shans Abari, Kashmir	Ordovician	O 7

Fossil No.	Fossil Name	Fossil Group	Formation and area	Formation Age	Formation Number
L 9	Lenticulina kossmatina	Foraminifera	Karai Shale Formation and Maravattur Clay Formation	Cretaceous	K 26 M 28
L 10	Lenticulina macrodisca	Foraminifera	Kallakkudi Limestone Formation	Cretaceous	K 11
L 11	Lenticulina sp.	Foraminifera	Chikkim Formation, Himachal Pradesh	Cretaceous	C 31
L 12	Lepidocorbitoides	NA	Khalsi Formation, J & K	Cretaceous	K 42
L 13	Lepidocyclina	Foraminifera	Kand Formation, Gujarat Mainland	Miocene	K 19
L 14	Lepidocyclina sumatraensis		Khari Series, Kutch		K 44
L 15	Lepidodendropsis sggillariode	Fern	Gund Formation, Kashmir	Lr. Carboniferous	G 31
L 16	Lepidorbitoides sp.	Foraminifera	Kallankurichi Formation	Cretaceous	K 13
L 17	Lepidosiggilaria quadrata	Plantae	Gund Formation, Kashmir	Lr. Carboniferous	G 31
L 18	Lepidostrobus kashmirensis	Fern	Gangamopteris Beds, Kashmir	Permian	G 5
L 19	Lepidotus	Fish	Takli Formation, Nagpur	Cretaceous-Paleocene	T 4
L 20	Lepisosteus indicus	Fish	Inter-trappean beds of Asifabad, Naskal and Rangapur, Andhra Pradesh; Anjar and Kora, Gujarat; Kisalpuri, Madhya Pradesh; Nagpur, Maharashtra	Cretaceous	i 8- 1a,c,d; 2a,c; 4a; 5b
L 21	Leptaena	Brachiopoda	Ordovician -Silurian rocks of Shans Abari, Kashmir	Ordovician	O 7
L 22	Leptelloidea	Brachiopoda	Ordovician -Silurian rocks of Shans Abari, Kashmir, Kashmir	Ordovician	O 7
L 23	Leptodactylidae indet	Amphibian	Inter-trappean beds of Naskal, Andhra Pradesh	Cretaceous	i 8- 1c
L 24	Lewesiceras vaju	Ammonite	Kulakkalnattam Formation	Cretaceous	K 69
L 25	Lima melaancholica	Lamellibranchia	Spiti Formation	Up. Jurassic to Lr. Cretaceous	S 61
L 26	Lima tristicula				

Fossil No.	Fossil Name	Fossil Group	Formation and area	Formation Age	Formation Number
L 27	Lingulella sp	Brachiopoda	Tal Formation, Himachal Pradesh	Cambrian	T 5
L 28	Lingulella spitiensis		Tethyan Basin, Himalaya		T 20
L 29	Lingullela haimantenensis		Kunzun La Formation, Spit		K 73a
L 30	Linmocypridea eephymatos	Ostracoda	Inter-trappean beds of Anjar and Kora, Gujarat	Cretaceous	i 8- 2a,c
L 31	Linoproductus	Brachiopoda	Saser Brangsa Formation	Carboniferous	S 13
			Permian rocks, Niti, Malla Johar and Byan areas, Uttarakhand	Permian	P 31
L 32	Linoproductus cora		Miri Formation, Arunachal Pradesh		M 54
L 33	Lisania	Trilobita	Kunzun La Formation, Zanskar, J&K	Cambrian	K 73b
L 34	Lissemy punctata	Reptilia (Trutle)	Quaternary sediments of Tamil Nadu, Soyamali	Pleistocene	Q 9
L 35	Lissochonetes carbonifera	Brachiopoda	Miri Formation, Arunachal Pradesh	Permian	M 54
L 36	Listriodon	Suidae	Siwalik Group, Lower Division	Miocene	S 51a
L 37	Lithistida pharetronids	Porifera	Kallakkudi Limestone Formation	Cretaceous	K 11
L 38	Lithoparaphysis paproensis	PF	Inter-trappean beds of Lalitpur, Uttar Pradesh	Cretaceous	I 8- 7a
L 39	Litriodon	Suidae	Siwalik Group, Lower Division	Miocene	S 51a
L 40	Lobothyria	Ammonite	Umia Stage, Kutch, Gujarat	Cretaceous	U 11
L 41	Lobtannularia ensifolia	Plantae	Mamal Formation, J & K	Lr. Permian	M 19
L 42	Lonsdaleia salinaria	Conodonta	Tethyan Basin, Himalaya	Permian	T 20
L 43	Lopha	Lamellibranchia	Ariyalur Group, Trichinopoly	Cretaceous	A 35
L 44	Lorquatisphisctes labyensis	Plantae (a fruit tree)	Bhadasar and Baisakhi Formations, Jaisalmer, Rajasthan	Jurassic	B 42 &B 14
L 45	Lorquatisphisctes rebillyi				
L 46	Lorquatisphisctes tenuistriata				

Fossil No.	Fossil Name	Fossil Group	Formation and area	Formation Age	Formation Number
L 47	Loxonema	Gastropoda	Yamne Formation, Arunachal Pradesh	Permian	Y 2
L 48	Loxostrmum cf. limbatum	Foraminifera	Miliolite / Porbandar Stone, Gujarat	Pleistocene	M 49
L 49	Ludbrookia sp.	Lamellibranchia	Khalsi Formation, Jammu and Kashmir	Cretaceous	K 42
L 50	Lunatisporites	PF	Gondwana Supergroup, Lower Division	Permian to Triassic	S 51a
L 51	Lunulites	Bryozoa	Tirupati Formation	Lr. Cretaceous	T 34
L 52	Lycopodiumsporit es sp.	Fern	Umia Formation, Kutch	Cretaceous	U 8
			Raghvapuram Formation		R 2
L 53	Lycyaena	Carnivora	Siwalik Group, Middle Division	Pliocene	S 51b
L 54	Lymnaea oviformes	Gastropoda	Infra-trappean (=Lameta) beds of Dongargaon, Maharashtra	Cretaceous	i 5-4a
L 55	Lymnaea sp.		Inter-trappean beds of Rajahmundry, Andhra Pradesh; Kora and Lakshmipur, Gujarat; Biligi, Gurmatkal and Upparhatti, Karnataka; Mandla and Sagar, Madhya Pradesh	Cretaceous	i 8- 1b; 2c,d; 3a,b,c; 4d,e
L 56	Lymnaea subulata		Infra-trappean (=Lameta) beds of Jabalpur, Madhya Pradesh		i 5-3a
L 57	Lyria Formosa	Gastropoda	Niniyur Group/Formation	Early Paleocene	N 23
L 58	Lyriapsis	Trilobita	Kunzun La Formation, Zanskar, J&K	Cambrian	K 73b
M 1	Macrocephalites	Cephalopoda	Lapthal Member, Uttarakhand	Jurassic	L 13
M 2	Macrocephalites	Cephalopoda	Macrocephalus beds, Kutch	Jurassic	M 2
			Pachcham coral bed, Kutch		P 2
			Pachcham shell Limestone, Kutch		P 4

Fossil No.	Fossil Name	Fossil Group	Formation and area	Formation Age	Formation Number
M 3	Madtsoiinae indet	Reptilia	Inter-trappean beds of Naskal, Andhra Pradesh	Cretaceous	i 8- 1a
M 4	Maldeotaia bandalica	Invertebrata (Annelida)	Tethyan Basin, Himalaya	Cambrian	T 20
M 5	Malonodium	Plantae	Umia Beds, Kutch, Gujarat	Cretaceous	U 7
M 6	Mammites cancellatum	Ammonite	Karai Shale Formation and Maravattur Clay Formation	Cretaceous	K 26 M 28
M 7	Mangolianella	Ostracoda	Inter-trappean beds of Asifabad, Andhra Pradesh & Nagpur, Maharashtra	Cretaceous	i 8- 1a & 5b
M 8	Mangolianella cylindrica		Inter-trappean beds of Kora & Lakshmipur, Gujarat & Gurmatkal, Karnataka		i 8-2c,d &3b
M 9	Marattiopsis macrocarpa	Fern	Tirupati Formation	Lr. Cretaceous	T 34
M 10	Marginifera	Brachiopoda	Boulder Slate Formation	Permian	B 71
M 11	Marginifera himalayensis		Saser Brangsa Formation	Carboniferous	S 13
			Gungri Formation	Permian	G 33
			Productus Shale, Spiti	Permian	P 51
M 12	Marginotruncana indica	Foraminifera	Uttatur Group, Trichinopoly	Cretaceous	U 21
M 13	Marginotruncana indica	Foraminifera	Karai Shale Formation and Maravattur Clay Formation	Cretaceous	K 26 M 28
M 14	Mastodonosaurus indicus	Labyrintho-dontia	Pachmarhi Formation	Triassic	P 5
M 15	Matonidium geopperiti	Plantae (Angiospermae)	Dharangadhara Group, Gujarat	Cretaceous	D 28
M 16	Mayaites	Ammonite	Dhosa Oolite, Kutch	Jurassic	D 34
M 17	Mayites waagens		Spiti Formation/Shale	Up. Jurassic to Lr. Cretaceous	S 61
M 18	Maytilus	Lamellibranchia	Vemavaram Formation	Cretaceous	V 7
M 19	Meagalosaurus	Dinosauria	Takli Formation, Nagpur	Cretaceous-Paleocene	T 4
M 20	Meekoceras lilangense	Cephalopoda	Meekoceras Zone, Spiti	Triassic	M 39

Fossil No.	Fossil Name	Fossil Group	Formation and area	Formation Age	Formation Number
M 21	Meekoceras markhami		Rambakot Member, Uttarakhand		R 13
M 22	Meekoceras varaha		Meekoceras Zone, Spiti		M 39
			Rambakot Member		R 13
M 23	Meekospira	Gastropoda	Yamne Formation, Arunachal Pradesh	Permian	Y 2
M 24	Megadesmus gracilis	Lamellibranchia	Agglomeratic Slate, J & K	Carboniferous-Permian	A 4
M 25	Megalodon ladakhensis	Cephalopoda	Megalodon Limestone, Spiti	Triassic	M 40
M 26	Megalodontoides	Insecta	Tagling Limestone, Spiti	Jurassic	T 1
M 27	Megaloolithus baghensis	Dinosauria	Inter-trappean beds of Kisalpuri, Madhya Pradesh	Cretaceous	i 8- 4a
M 28	Megalosaurus	Reptilia	Kallamedu Formation, Ariyalur Gr, Trichinopoly	Cretaceous	K 12
M 29	Melania hunter	Gastropoda	Infra-trappean (=Lameta) beds of Dongargaon, Maharashtra	Cretaceous	i 5-4a
M 30	Meretrix casta	Lamellibranchia	Chilka Lake deposit of estuarine shells, Orissa	pleistocene	C 33
M 31	Mesembrixylon	Plantae (tree trunk)	Pariwar Formation, jaisalmer, Rajasthan	Cretaceous	P 23
M 32	Mesoorbitolina	Foraminifera	Hundri Formation, Jammu and Kashmir	Cretaceous	H 21
M 33	Mespuzosia indopacificum	Ammonite	Kulakkalnattam Formation, Trichin-opoly Gr, Trichinopoly	Cretaceous	K 69
M 34	Metapolygnathus polygnathisformis	Conodonta	Triassic of NW Himalaya, Kashmir, Ladakh, Spiti	Up. Triassic	T 40
M 35	Metaposaurus	Labyrinthodontia	Pachmarhi Formation	Triassic	P 5
M 36	Michelina salinaria	Invertebrata	Kukti Formation, Lahaul, Himachal Pradesh	Permian	K 68
M 37	Micracanthoceras	Ammonite	Umia ammonite bed, Kutch	Cretaceous	U 6
M 38	Microcachryidites	PF	Umia Formation, Kutch	Cretaceous	U 8
M 39	Microcachryidites antarcticus	PF	Raghvapuram Formation	Lr. Cretaceous	R 2

Fossil No.	Fossil Name	Fossil Group	Formation and area	Formation Age	Formation Number
M 40	Microchara grambasti	Charophyta (Algae)	Infra-trappean (=Lameta) beds of Dongargaon, Maharashtra	Cretaceous	i 5-4a
M 41	Microchara sp.	Charophyta (Algae)	Inter-trappean beds of Gurmatkal, Karnataka	Cretaceous	i 8- 3b
M 42	Microrhabdulus decorates	NF	Karai Shale Formation and Maravattur Clay Formation	Cretaceous	K 26 M 28
M 43	Microsolena	Coral	Kallakkudi Limestone Formation	Cretaceous	K 11
M 44	Miliammina	Foraminifera	Surma Group, NE India	Early Miocene	S 73
M 45	Miliolids	Foraminifera	Aida Stage, Kutch	Oligocene	A 6
M 46	Miliolina Genus	Foraminifera	Miliolite / Porbandar Stone, Gujarat	Pleistocene	M 49
M 47	Miogypsina	Foraminifera	Kand Formation, Gujarat Mainland	Lr. Miocene	K 19
			Gaj Formation,Gujarat		G 1
M 48	Modiola lidarensis	Lamellibranchia	Salooni Formation, Lahaul, Himachal Pradesh	Permian	S 4
			Tethyan Basin, Himalaya	Carboniferous	T 20
M 49	Mongolianella	Ostracoda	Inter-trappean beds, Mumbai	Cretaceous	i 8-5a
M 50	Mongolianella cylindrical		Inter-trappean beds of Kora and Lakshmipur, Gujarat; Gurmatkal, Karnataka		i 8-2c,d; 3b
M 51	Mongolianella khamariniensis		Infra-trappean (=Lameta) beds of Rahioli, Gujarat		i 5-2a
M 52	Mongolianella palmosa		Infra-trappean (=Lameta) beds of Jabalpur, Madhya Pradesh		i 5-3a
M 53	Mongolianella sp.		Infra-trappean (=Lameta) beds of Dongargaon, Maharashtra		i 5-4a
M 54	Mongolocypris cf. M. gigantean	Ostracoda	Infra-trappean (=Lameta) beds of Rahioli, Gujarat	Cretaceous	i 5-2a

Fossil No.	Fossil Name	Fossil Group	Formation and area	Formation Age	Formation Number
M 55	Monomorphichnus	TF	Thango Formation, Kinnaur, Himachal Pradesh	Ordovician	T 23
			Kunzun La Formation, Spiti	Lr. Cambrian	K 73a
M 56	Monotis salinaria	Lamellibranchia	Monotis Shales, Spiti	Triassic	M 57
M 57	Montlivaltia	Actinozoa	Pachcham coral bed, Kutch	Jurassic	P 2
M 58	Mourlonia	Gastropoda	Yamne Formation, Arunachal Pradesh	Permian	Y 2
M 59	Murospora	PF	Umia Formation, Kutch	Cretaceous	U 8
M 60	Mysidioptera	Lamellibranchia	Yamne Formation, Arunachal Pradesh	Permian	Y 2
M 61	Mytiloides	Lamellibranchia	Bagh Beds, Narmada Basin	Cretaceous	B 8
N 1	Namadoceras scindia	Ammonite	Bagh Beds, Narmada Basin	Cretaceous	B 8
N 2	Namodoceras bosai				
N 3	Nautilus	Cephalopoda	Tirupati Formation	Cretaceous	T 34
			Vemavaram Formation		V 7
			Kulakkalnattam Formation		K 69
N 4	Nautilus danicus		Niniyur Group/Formation	Early Paleocene	N 23
N 5	Neilsonia	Gastropoda	Yamne Formation, Arunachal Pradesh	Permian	Y 2
N 6	Nemegtichara grambasti	Charophyta (Algae)	Inter-trappean beds of Rangapur, Andhra Pradesh; Kora, Gujarat; Gurmatkal, Karnataka; Mamoni, Rajasthan	Cretaceous	i 8- 1d; 2c; 3b; 6a
			Infra-trappean (=Lameta) beds of Dongargaon, Maharashtra		i 5-4a
N 7	Nemigtichara		Takli Formation, Nagpur	Cretaceous-Paleocene	T 4
N 8	Neochonets	Brachiopoda	Yamne Formation, Arunachal Pradesh	Permian	Y 2

Fossil No.	Fossil Name	Fossil Group	Formation and area	Formation Age	Formation Number
N 9	Neocomites	Ammonite	Spiti Formation/ Shale, Malla Johar Supergroup	Lr. Cretaceous	S 61
			Sancha Malla Group, Uttarakhand	Up. Jurassic to Cretaceous	S 5
			Giumal Formation	Cretaceous	G 15
N 10	Neocosmoceras	Ammonite	Spiti Formation/Shale, Malla Johar Supergroup	Lr. Cretaceous	S 61
			Sancha Malla Group, Uttarakhand	Up. Jurassic to Cretaceous	S 5
N 11	Neocosmoceras gr. octagonus		Giumal Formation	Cretaceous	G 15
N 12	Neogondolella carinata	Conodonta	Triassic of NW Himalaya, Spiti	Lr. Triassic	T 40
N 13	Neogondolella constriata		Triassic of NW Himalaya, Kiogarh Valley	Mid. Triassic	T 40
N 14	Neogondolella mombergensis				
N 15	Neogondolella navicula				
N 16	Neogondolella subcarinata		Khunamuh Formation, J &K	Permian-Triassic	K 51
			Zewan Formation, J & K	Permian	Z 4
N 17	Neonereites	TF	Kunzun La Formation, Spiti	Lr. Cambrian	K 73a
N 18	Neoprioniodus confluence	Conodonta	Muth Formation, Kashmir, Kishtwar, Ladakh and Kumaon	Devonian	M 66
			Tethyan Basin, Himalaya	Carboniferous	T 20
			Syringothyris Lst. Formation, Kashmir	Carboniferous	S 79
N 19	Neospathodus lanceolatus	Conodont	Triassic of NW Himalaya, Niti Pass, Uttarakhand	Up. Triassic	T 40
N 20	Neospathodus timorensis		Triassic of NW Himalaya, Spiti	Lr.Triassic	T 40
N 21	Neospirifer	Brachiopoda	Gechang Formation	Permian	G 10
N 22	Neospirifer		Boulder Slate Formation		B 71

Fossil No.	Fossil Name	Fossil Group	Formation and area	Formation Age	Formation Number
N 23	Neospirifer		Yamne Formation, Arunachal Pradesh		Y 2
N 24	Neospirifer banbadhurensis		Lipak Formation	Carboniferous	L 24
N 25	Neospirifer cf. fasciger (Keyserling)		Seti Khola Formation, Bhutan	Permian	S 25
N 26	Nephrolepidina	Foraminifera	Gaj Formation, Gujarat	Lr. Miocene	G 1
N 27	Nerinea	Gastropoda	Niniyur Formation, Trichinopoly	Paleocne	N 23
N 28	Nesokia	Rodentia	Siwalik Group, Upper Division	Pleistocene	S 51c
N 29	Nilssonia morrisiana	Plantae (Spermatophyta)	Tirupati Formation	Lr. Cretaceous	T 34
N 30	Nodosaria	Foraminifera	Quaternary sediments of Gujarat Mainland	Quaternary	Q 5
			Chikkim Formation	Cretaceous	C 31
N 31	Nodosaria zippei		Inter-trappean beds of Rajahmundry, Andhra Pradesh	Cretaceous	i 8- 1b
N 32	Noeggerothiopsis	Cordaitales	Chhongtash Formation, J&K	Permian	C 27
N 33	Noeggerrathiopsis hislopi		Gondwana Supergroup, Lower Division	Permian to Triassic	G 23a
N 34	Nonion scapha	Foraminifera	Miliolite / Porbandar Stone, Gujarat	Pleistocene	M 49
N 35	Nonion sp. Indet		Inter-trappean beds of Rajahmundry, Andhra Pradesh	Cretaceous	i 8- 1b
N 36	Nonionella	Foraminifera	Broach Formation, Gujarat	Pliocene	B 72
			Quaternary sediments of Gujarat mainland	Quaternary	Q 5
N 37	Notopteridarum nolfi	Fish	Inter-trappean beds of Naskal and Rangapur, Andhra Pradesh	Cretaceous	i 8- 1c,d
N 38	Notopterus	Fish	Takli Formation, Nagpur	Cretaceous-Paleocene	T 4
N 39	Nucleolina diluta	Ostracoda	Inter-trappean beds of Rajahmundry, Andhra Pradesh	Cretaceous	i 8- 1b
N 40	Nummulites atacicus	Foraminifera	Subathu Formation, Uttarakhand	Eocene to Miocene	S 65

Fossil No.	Fossil Name	Fossil Group	Formation and area	Formation Age	Formation Number
N 41	Nummulites cf. globulus				
N 42	Nummulites fabianii		Baratang Group, Andaman and Nicobar Islands	Cretaceous	B 25
N 43	Nummulites sps.		Dinod Formation, Gujarat	Eocene	D 40
O 1	Obolella sp.	Brachiopoda	Tal Formation, Himachal Pradesh	Lr. Cambrian	T 5
O 2	Odontodiscoceras gr. Odontodiscus	Ammonite	Giumal Formation	Cretaceous	G 15
O 3	Oldhamina sp.	Brachiopoda	Miri Formation, Arunachal Pradesh	Permian	M 54
O 4	Olivooides multisulcatus	Invertebrata	Tethyan Basin, Himalaya	Cambrian	T 20
O 5	Oneotodus spp.	Conodonta	Kunzun La Formation, Spiti	Cambrian	K 73a
O 6	Onpehiopsis psilotoides	Plantae	Dharangadhara Group, Gujarat	Cretaceous	D 28
O 7	Onychiopsis	Fern	Umia Beds, Kutch, Gujarat	Cretaceous	U 7
			Intertrappean, Dubrajpur Formation, Rajmahal Basin	Triassic	i 8-9
O 8	Operculina cf. patalensis	Foraminifera	Subathu Formation, Uttarakhand	Eocene to Miocene	S 65
O 9	Operculina sp.		Waior Stage, Kutch	Oligocene	W 2
			Ramania Stage, Kutch	Oligocene	R 12
O 10	Operculinoides sp.	Foraminifera	Ramania Stage, Kutch	Oligocene	R 12
O 11	Ophiceras sakuntala	Cephalopoda	Ophiceras Zone, Spiti	Triassic	O 6
			Rambakot Member, Uttarakhand		R 13
O 12	Ophimorpha	TF	Ottakovil Formation	Cretaceous	O 9
O 13	Orbitoides	Foraminifera	Mahdek Formation, Assam/Meghalaya	Cretaceous	M 4
O 14	Orbitolina	Foraminifera	Dras Formation	Cretaceous	D 48
			Hundri Formation, J & K		H 21
O 15	Orbitolina discoidea		Khalsi Formation, J & K	Cretaceous	K 42
O 16	Orbitolina Kashmirica				

Fossil No.	Fossil Name	Fossil Group	Formation and area	Formation Age	Formation Number
O 17	Orbitolina parma				
O 18	Orbitolina piteus				
O 19	Orbulina cf. O universa	Foraminifera	Inter-trappean beds of Rajahmundry, Andhra Pradesh	Cretaceous	i 8- 1b
O 20	Orionoides	Cephalopoda	Athleta Beds, Kutch	Jurassic	A 42
O 21	Ornithoid eggshell fragments	Dinosauria	Inter-trappean beds of Anjar, Gujarat	Cretaceous	i 8- 2a
O 22	Orthis	Brachiopoda	Ordovician-Silurian rocks of Shans Abani Syncline, Handwar, J & K	Ordovician	O 7
			Takche Formation, Spiti	Silurian-Ordovician	T 3
			Shiala Formation, Uttarakhand	Ordovician	S 31
O 23	Orthis (Dlmenella) bouchordi		Variegated Formation	Silurian-Ordovician	V 5
O 24	Orthis aff. Bistriata		Muth Formation, Byans, Uttarakhand	Devonian	M 66
O 25	Orthis aff. Spitiensis		Tethyan Basin, Himalaya	Silurian	T 20
O 26	Orthis rustica		Muth Formation, Spiti	Devonian	M 66
O 27	Orthoceras	Cephalopoda	Muth Formation, Byans, Uttarakhand	Devonian	M 66
O 28	Orthoceras commutatum		Tethyan Basin, Himalaya	Ordovician	T 20
O 29	Orthothetis	Brachiopoda	Shiala Formation, Uttarakhand	Ordovician	S 31
			Lipak Formation	Carboniferous	L 24
O 30	Orthothetis aff. Crenistria		Muth Formation, Spiti	Devonian	M 66
O 31	Oryctocephalus	Trilobita	Kunzun La Formation, Spit	Mid. Cambrian	K 73a
O 32	Oryctocephalus indicus		Tethyan Basin, Himalaya	Cambrian	T 20
O 33	Oryctocephalus salteri		Tethyan Basin, Himalaya	Cambrian	T 20
O 34	Osteoglossidarum intertrappus	Fish	Inter-trappean beds of Naskal and Rangapur, Andhra Pradesh	Cretaceous	i 8- 1c,d
O 35	Osteoglossum	Fish	Takli Formation, Nagpur	Cretaceous-Paleocene	T 4

Fossil No.	Fossil Name	Fossil Group	Formation and area	Formation Age	Formation Number
O 36	Ostrea	Lamellibranchia	Kand Formation, Gujarat Mainland	Lr. Miocene	K 19
			Coastal Gondwana Formations, Ramnath-puram, Tamil Nadu	Lr.Cretaceous	C 45
			Giumal Formation	Cretaceous	G 15
			Subathu Formation, Uttarakhand	Eocene to Miocene	S 65
O 37	Otozamites imbricatus	Cycadophyta	Pariwar Formation, jaisalmer	Cretaceous	P 23
			Raghvapuram Formation	Lr. Cretaceous	R 2
O 38	Otoceras clirei	Cephalopoda	Otoceras Zone, Spiti	Triassic	O 8
O 39	Otoceras woodwardi		Otoceras Zone, Spiti	Triassic	O 8
			Rambakot Member		R 13
O 40	Ovocytheridea raoi	Ostracoda	Inter-trappean beds of Rajahmundry, Andhra Pradesh	Cretaceous	i 8- 1b
O 41	Oxytropidoceras (Micropelia) douglasi	Cephalopoda	Dras Formation, Jammu and Kashmir	Cretaceous	D 48
O 42	Oysters	Lamellibranchia	Uttatur Group, Trichinopoly	Cretaceous	U 21
			Kallakkudi Limestone Formation		K 11
O 43	Ozarkodina macer	Conodonta	Syringothyris Limestone Formation, Kashmir	Carboniferous	S 79
P 1	Pacenticeras tamulicum	Ammonite	Anaipadi Formation	Cretaceous	A 22
P 2	Pachydiscus otokodensis	Ammonite	Ottakovil Formation	Cretaceous	O 9
P 3	Pachydiscus egertoni		Kallankurichi Formation		K 13
P 4	Deleted		-----		----
P 5	Pachysphinctes granti	Ammonite	Spiti Formation	Jurassic	S 61
P 6	Pachysphinctes linguiferus				
P 7	Pagetia greibachis	Trilobita	Parahio Series, Spiti	Mid.Cambrian	P 21
P 8	Pagiophyllum	Coniferales	Gangapur Formation.	Lr. Cretaceous	G 6

Fossil No.	Fossil Name	Fossil Group	Formation and area	Formation Age	Formation Number
			Intertrappean, Dubrajpur Formation, Rajmahal Basin	Triassic	i 8-9
P 9	Palaeindicus	Crocodylidae	Pleistocene sediments of Kurnool, Ghod, Manjra, Pravara and Godavari	Pleistocene	P 37a
			Pleistocene sediments of Central Narmada (upper group), Palmar		P 37b
P 10	Palaeophycus	TF	Subathu Formation, Uttarakhand	Eocene to Miocene	S 65
P 11	Palaeopristolepis chiplonkari	Fish	Inter-trappean beds of Bamanbor, Gujarat	Cretaceous	i 8- 2b
P 12	Palaeopristolepis feddeni				
P 13	Paleoniscus		Gangamopteris Beds	Permian	G 5
P 14	Palmoconcha rajui	Ostracoda	Inter-trappean beds of Rajahmundry, Andhra Pradesh	Cretaceous	i 8- 1b
P 15	Palmoconcha rajuvi	Ostracoda	Infra-trappean beds, Rajahmundry	Cretaceous	i 5-1b
P 16	Palmoxylan	Palm	Bhuj Beds, Kutch	Cretaceous	B 53
			Baripada Beds, Orissa	Miocene	B 26
P 17	Palmoxylon schenk		Infra-trappean beds, Jabalpur	Cretaceous-Paleocene	i 5-3a
P 18	Palmula rugosa	Foraminifera	Chikkim Formation	Cretaceous	C 31
P 19	Palorbitolina	Foraminifera	Hundri Formation, Jammu and Kashmir	Cretaceous	H 21
P 20	Paludina deccanesis	Gastropoda	Infra-trappean (=Lameta) beds of Dongargaon, Maharashtra	Cretaceous	i 5-4a
P 21	Paludina hislopi		Infra-trappean (=Lameta) beds of Jabalpur, Madhya Pradesh; Dongargaon, Maharashtra	Cretaceous	i 5-3a; 4a
P 22	Paludina normalis		Infra-trappean (=Lameta) beds of Jabalpur, Madhya Pradesh	Cretaceous	i 5-3a

Fossil No.	Fossil Name	Fossil Group	Formation and area	Formation Age	Formation Number
P 23	Paludina pyramis		Infra-trappean (=Lameta) beds of Jabalpur, Madhya Pradesh	Cretaceous	i 5-3a
P 24	Paludina sp.		Inter-trappean beds of Rajahmundry, Andhra Pradesh; Anjar and Lakshmipur, Gujarat; Gurmatkal and Upparhatti, Karnataka; Mandla and Sagar, Madhya Pradesh	Cretaceous	i 8- 1b; 2a,d; 3b,c; 4d,e
P 25	Panchyspintes	Invertebrata	Middle Katrol, Kutch	Jurassic	M 45
P 26	Panderodus	Conodonta	Takche Formation, Spiti	Silurian-Ordovician	T 3
P 27	Papio	Primates	Siwalik Group, Upper Division	Pleistocene	S 51c
P 28	Paraboliceras criptoptychus	Ammonite	Spiti Formation	Jurassic	S 61
P 29	Paraboliceras griesbachi		Spiti Formation		S 61
P 30	Paraboliceras himalayanus		Spiti Formation		S 61
P 31	Paraboliceras mutilis		Spiti Formation		S 61
P 32	Paraboliceras polysphinctus		Spiti Formation		S 61
P 33	Paraboliceras rectacostatus		Spiti Formation		S 61
P 34	Paraboliceras suineanus		Spiti Formation		S 61
P 35	Paraboliceras tibaticus		Spiti Formation		S 61
P 36	Paraboliceratoides sp.	Ammonite	Spiti Formation	Jurassic	S 61
P 37	Paracandona jabalpurensis	Ostracoda	Inter-trappean beds of Kisalpuri, Madhya Pradesh	Cretaceous	i 8- 4a
			Infra-trappean (=Lameta) beds of Jabalpur, Madhya Pradesh		i 5-3a
P 38	Paracenoceras	Cephalopoda	Dhosa Oolite, Kutch	Jurassic	D 34

Fossil No.	Fossil Name	Fossil Group	Formation and area	Formation Age	Formation Number
P 39	Parachaetetes aswapathi	Algae	Niniyur Formation, Trichinopoly	Paleocne	N 23
P 40	Paracypretta bhatiai	Ostracoda	Inter-trappean beds of Asifabad, Andhra Pradesh; Anjar and Kora, Gujarat; Gurmatkal, Karnataka; Kisalpuri and Mohgaon-Kalan, Madhya Pradesh	Cretaceous	i 8- 1a; 2a,c; 3b; 4a,b
P 41	Paracypretta elizabethae	Ostracoda	Inter-trappean beds of Kora and Lakshmipur, Gujarat; Gurmatkal, Karnataka	Cretaceous	i 8- 2c,d; 3b
P 42	Paracypretta jonesi	Ostracoda	Infra-trappean (=Lameta) beds of Dongargaon and Pisdura, Maharashtra	Cretaceous	i 5-4a,b
			Inter-trappean beds of Gurmatkal, Karnataka; Nagpur, Maharashtra		i 8- 3b; 5b
P 43	Parafusilina	Foraminifera	Khunamuh Formation, J &K	Permian	K 51
P 44	Paramarginifera himalayensis	Brachiopoda	Kuling Formation, Uttarakhand	Permian	K 69
P 45	Pararotalia mexicana	Foraminifera	Waior Stage, Kutch	Oligocene	W 2
P 46	Paraspenophyllum	Pteridophyta	Mamal Formation, J & K	Lr. Permian	M 19
P 47	Parastantanoceras	Ammoniteea	Bagh Beds, Narmada Basin	Cretaceous	B 8
P 48	Paratetralophodon	Proboscidea	Siwalik Group, Upper Division	Pleistocene	S 51c
P 49	Paratosaurus	Labyrinthodontia	Pachmarhi Formation	Triassic	P 5
P 50	Paratosuchus	Amphibian	Yerrapalli Formation	Triassic	Y 5
P 51	Pascoeites budavadensis	Ammonite	Coastal Gondwana Formations, Ramnathpuram, Tamil Nadu	Lr.Cretaceous	C 45
P 52	Pascoeites crassus		Coastal Gondwana Formations, Ramnathpuram, Tamil Nadu	Lr.Cretaceous	C 45
P 53	Pascoites crassus		Sriperumbudur Formation	Mid. Jurassic(?)	S 63

Fossil No.	Fossil Name	Fossil Group	Formation and area	Formation Age	Formation Number
P 54	Peckichara varians	Charophyta (Algae)	Inter-trappean beds of Kora, Gujarat; Gurmatkal, Karnataka; Nagpur, Maharashtra	Cretaceous	I 8- 2c; 3b; 5b
P 55	Pecopteris	Fern	Gangamopteris Beds, Kashmir	Permian	G 5
P 56	Pecten	Lamellibranchia	Uttatur Group, Trichinopoly	Cretaceous	U 21
			Fenestella Shale, Kashmir	Carboniferous	F 2
			Karai Shale Formation and Maravattur Clay Formation	Cretaceous	K 26 M 28
			Kallakkudi Limestone Formation	Cretaceous	K 11
			Kand Formation, Gujarat Mainland	Lr. Miocene	K 19
			Giumal Formation	Cretaceous	G 15
P 57	Pedium	TF	Kunzun La Formation, Spit	Lr. Cambrian	K 73a
P 58	Pelecycora sp.	Lamellibranchia	Subathu Formation, Uttarakhand	Eocene to Miocene	S 65
P 59	Peltoceratoides	Cephalopoda	Dhosa Oolite, Kutch	Jurassic	D 34
P 60	Peltoceras	Cephalopoda	Athleta beds, Chari Formation, Kutch	Jurassic	A 42
P 61	Pentacrinus	Echinoidea	Sillakudi Formation	Cretaceous	S 37
P 62	Pentamerus	Brachiopoda	Thango Formation, Kinnaur,Himachal Pradsh	Ordovicia	T 23
P 63	Pentamerus cf. sublinguifer	Brachiopoda	Muth Formation, Byans, Uttarakhand	Devonian	M 66
P 64	Pentoxylon sahnii	Gymnospermae	Gondwana Supergroup, Upper Division	Mid. Triassic-Lr. Cretaceous	G 23b
P 65	Perca cf. P. angusta	Fish	Inter-trappean beds of Bamanbor, Gujarat	Cretaceous	i 8- 2a
P 66	Percoideorum rangapurensis	Fish	Inter-trappean beds of Rangapur, Andhra Pradesh	Cretaceous	i 8- 1d
P 67	Percoideorum ellipticus		Inter-trappean beds of Rangapur, Andhra Pradesh		i 8- 1d
P 68	Percomorpha indet		Inter-trappean beds of Bamanbor, Gujarat	Cretaceous	i 8- 2b

Fossil No.	Fossil Name	Fossil Group	Formation and area	Formation Age	Formation Number
P 69	Perisphincites	Ammonite	Anceps Beds, Kutch	Jurassic	A 23
P 70	Peronopsis amplaxis	Trilobita	Parahio Series, Spiti	Mid. Cambrian	P 21
P 71	Persphinctes	Cephalopoda	Dhosa Oolite, Kutch	Jurassic	D 34
P 72	Phareodua sp.	Fish	Inter-trappean beds of Asifabad, Naskal and Rangapur, Andhra Pradesh; Kisalpuri, Madhya Pradesh; Nagpur, Mharashtra	Cretaceous	I 8- 1a,c,d; 4a; 5b
P 73	Phestia lyonensis	Lamellibranchia	Miri Formation, Arunachal Pradesh	Permian	M 54
P 74	Phillipsia cf. cliffordi	Trilobita	Lipak Formation	Carboniferous	L 24
P 75	Phoenicites brongniart	Palm	Infra-trappean beds of Jabalpur	Cretaceous-Paleocene	I 5-3a
P 76	Phoenocopsis	PF	Pachmarhi Formation	Triassic	P 5
P 77	Phragmothyrites eocaenica	PF	Inter-trappean beds of Lalitpur, Uttar Pradesh	Cretaceous	i 8- 7a
P 78	Phycode	TF	Kunzun La Formation, Spit	Lr. Cambrian	K 73a
P 79	Phycodes circinatum	TF	Tethyan Basin, Himalaya	Ordovician	T 20
P 80	Phycodes pedium		Tethyan Basin, Himalaya	Cambrian	T 20
P 81	Phycodes		Po Formation, Spiti-Kinnaur area	Carboniferous	P 38
P 82	Phygraca vesicularis	Lamellibranchia	Kallankurichi Formation	Cretaceous	K 13
P 83	Phyllotheca	Pteridophyta (Equisetales)	Gangamopteris Beds, Kashmir	Permian	G 5
			Bahreli Formation, Arunachal Pradesh		B 10
P 84	Phyloceras	Cephalopoda	Gajansar Beds, Kutch	Jurassic	G 2
P 85	Physa	Gastropoda	Trichinopoly Group, Trichinopoly	Cretaceous	T 44
P 86	Physa		Kulakkalnattam Formation	Cretaceous	K 69
P 87	Physa bridgerensis		Infra-trappean (=Lameta) beds of Dongargaon, Maharashtra	Cretaceous	i 5-4a

Fossil No.	Fossil Name	Fossil Group	Formation and area	Formation Age	Formation Number
P 88	Physa inflate		Infra-trappean (=Lameta) beds of Dongargaon, Maharashtra	Cretaceous	i 5-4a
P 89	Physa longiscula		Infra-trappean (=Lameta) beds of Dongargaon, Maharashtra	Cretaceous	i 5-4a
P 90	Physa prinsepii		Inter-trappean beds of Rajahmundry, Andhra Pradesh; Anjar, Kora and Lakshmipur, Gujarat; Biligi, Gurmatkal and Upparhatti, Karnataka; Padwar/Ranipur, Mandla and Sagar, Madhya Pradesh and Nagpur, Maharashtra	Cretaceous	i 8- 1b; 2a,c,d; 3a,b,c, 4c,d,e; 5b
P 91	Physa prinsepii		Infra-trappean (=Lameta) beds of Marepalli, Andhra Pradesh and Jabalpur, Madhya Pradesh; Dongargaon, Maharashtra	Cretaceous	i 5-1a;3a;4a
P 92	Pilobatidae indet	Amphibian	Inter-trappean beds of Nagpur, Maharashtra	Cretaceous	i 8- 5b
P 93	Placenticeras mintoi	Ammonite	Bagh Beds, Narmada Basin	Cretaceous	B 8
P 94	Placenticeras tamulicum		Kulakkalnattam Formation	Cretaceous	K 69
P 95	Plagiomus	TF	Kunzun La Formation, Spit	Lr. Cambrian	K 73a
P 96	Planolites	TF	Subathu Formation, Uttarakhand	Eocene to Miocene	S 65
P 97	Planolites		Thango Formation, Kinnaur,Himachal Pradesh	Oedovician	T 23
P 98	Planolites Nicholson		Bhuj Beds, Kutch, Gujarat	Cretaceous	B 53
P 99	Planolites		Kunzun La Formation, Spiti	Lr. Cambrian	K 73a
P 100	Planolites		Garbyang Formation, Uttarakhand	Cambrian	G 8

Fossil No.	Fossil Name	Fossil Group	Formation and area	Formation Age	Formation Number
P 101	Planolites		Po Formation, Spiti-Kinnaur area	Carboniferous	P 38
P 102	Planomalina buxtorfi	Foraminifera	Karai Shale Formation and Maravattur Clay Formation	Cretaceous	K 26 M 28
P 103	Planorbis sp.	Gastropoda	Infra-trappean (=Lameta) beds of Jabalpur, Madhya Pradesh	Cretaceous	i 5-3a
P 104	Planulina bhatiai	Foraminifera	Infra-trappean (=Lameta) beds of Rajahmundry, Andhra Pradesh	Cretaceous	i 5-1b
P 105	Platychara compressa	Charophyta (Algae)	Inter-trappean beds of Kora, Gujarat; Gurmatkal, Karnataka	Cretaceous	i 8- 2c; 3b
P 106	Platychara perlata		Takli Formation, Nagpur	Cretaceous-Paleocene	T 4
			Inter-trappean beds of Asifabad, Rangapur, Andhra Pradesh; Kora, Gujarat; Gurmatkal, Karnataka; Nagpur, Maharashtra	Cretaceous	i 8- 1a,d; 2c; 3b; 5b
P 107	Platychara rajamundrica		Inter-trappean beds of Rajahmundry, Andhra Pradesh; Kora, Gujarat	Cretaceous	i 8- 1b; 2c
P 108	Platychara raoi		Inter-trappean beds of Kora, Gujarat; Gurmatkal, Karnataka	Cretaceous	i 8- 2c; 3b
P 109	Platychara sahnii		Inter-trappean beds of Rajahmundry, Andhra Pradesh; Kora, Gujarat; Gurmatkal, Karnataka; Nagpur, Maharashtra	Cretaceous	i 8- 1b; 2c; 3b; 5b
P 110	Platyeichum brenensis (Reed)	Gastropoda	Miri Formation, Arunachal Pradesh	Permian	M 54
P 111	Platyteichum sp.		Seti Khola Formation, Bhutan		S 25
P 112	Plaziogmus	TF	Tethyan Basin, Himalaya	Late Precambrian	T 20
P 113	Pleuratomaria	Gastropoda	Po Formation, Spiti-Kinnaur area	Carboniferous	P 38
			Miri Formation, Arunachal Pradesh	Permian	M 54

Fossil No.	Fossil Name	Fossil Group	Formation and area	Formation Age	Formation Number
P 114	Pleurotomaria umariensis		Umaria Marine Bed and Manendragarh Marine Bed	Permian	U 5 & M 22
P 115	Pleurocephalites	Cephalopoda	Macrocephalus beds, Kutch	Jurassic	M 2
P 116	Plicatipollenites	PF	Lapthal Member, Uttarakhand	Jurassic	L 13
			Spiti Member, Uttarakhand	Up. Jurassic	S 62
			Kuling Shale Formation, Uttarakhand	Permian	K 70
P 117	Plochmopeltinites cooksoni	PF	Inter-trappean beds of Lalitpur, Uttar Pradesh	Cretaceous	i 8- 7a
P 118	Plycodes	TF	Garbyang Formation, Uttarakhand	Cambrian	G 8
P 119	Podocarpaceae	Coniferales	Cuddalore Sandstone Formation,	Miocene	C 56
P 120	Podosporites	PF	Chari Formation and Katrol Formation, Kutch	Jurassic	C 16 & K 38
			Umia Formation, Kutch	Cretaceous	U 8
P 121	Podozomites lanceolatus	Plantae	Tirupati Formation	Lr. Cretaceous	T 34
			Sivaganga Beds, Tamil Nadu	Lr. Cretaceous	S 49
P 122	Polyalthioxylon bande	Plantae (Fossil Wood)	Infra-trappean beds, Jabalpur	Cretaceous-Paleocene	i 5-3a
P 123	Polygnathus commuis communis	Conodonta	Syringothyris Lst. Formation, Kashmir	Carboniferous	S 79
P 124	Polypora ampla	Bryozoa	Badhaura Formation, Bikaner-Nagaur Basin, Rajasthan	Permian	B 5
P 125	Polypora megastoma		Miri Formation, Arunachal Pradesh	Permian	M 54
P 126	Polypora sp.		Khunamuh Formation, J &K		K 51
P 127	Porypora middlemissi		Boulder Slate Formation		B 71
P 128	Potamides angulatus	Gastropoda	Chilka Lake deposit of estuarine shells, Orissa	pleistocene	C 33
P 129	Praeorbitolina	Foraminifera	Hundri Formation, Jammu and Kashmir	Cretaceous	H 21
P 130	Praeundulomya	Lamellibranchia	Agglomeratic Slate Formation, J & K	Carboniferous-	A 4

Fossil No.	Fossil Name	Fossil Group	Formation and area	Formation Age	Formation Number
P 131	Prinodon ambadongerensis	Fish (Shark teeth)	Bagh Beds (Amba Dongar) Narmada Basin	Cretaceous	B 8
P 132	Priondomya	NA (a marine fauna)	Talchir Formation, Gondwana SG	Permian	T 6
P 133	Procerites	Cephalopoda	Pachcham Coral Bed, Kutch	Jurassic	P 2
P 134	Proclydonautilus griesbachi	Cephalopoda	Kioto Member, Uttarakhand	Triassic	K 55
P 135	Productella	Brachiopoda	Muth Formation, Spiti	Devonian	M 66
			Lipak Formation	Carboniferous	L 24
P 136	Productus	Brachiopoda	Syringothyris Lst. Formation, Kashmir	Carboniferous	S 79
			Fenestella Shale, Kashmir		F 2
			Lipak Formation		L 24
			Boulder Slate Formation	Permian	B 71
			Yamne Formation, Arunachal Pradesh	Permian	Y 2
			Umaria Marine Bed and Manendragarh Marine Bed	Permian	U 5 & M 22
			Salooni Formation, Lahaul, Himachal Pradesh	Permian	S 4
P 137	Productus gangeticus		Gungri Formation, Himachal Pradesh	Permian	G 33
P 138	Productus lineatus		Po Formation, Spiti-Kinnaur, Himachal Pradesh	Carbonni-ferous	P 38
P 139	Productus sabriculus				
P 140	Prognoblattina Columbiana	Insecta	Gangamopteris Beds	Permian	G 5
P 141	Prograyiceras	Cephalopoda	Kanthkot Sandstone, Kutch	Jurassic	K 23
P 142	Prograyiceras grayi		Spiti Formation	Up. Jurassic to Lr. Cretaceous	S 61
P 143	Propalaeomeryx	Giraffidae	Siwalik Group, Lower Division	Miocene	S 51a
P 144	Proplacenticeras	Ammonite	Bagh Beds, Narmada Basin	Cretaceous	B 8

Fossil No.	Fossil Name	Fossil Group	Formation and area	Formation Age	Formation Number
P 145	Propora himalaica	Actinozoa	Thango Formation, Kinnaur, Himachal Pradesh	Ordovician	T 23
P 146	Propora himalaica	Actinozoa	Takche Formation, Spiti	Ordovician-Silurian	T 3
P 147	Propora (Lyella) americana		Thango Formation, Kinnaur, Himachal Pradesh	Ordovician	T 23
			Takche Formation, Spiti	Ordovician-Silurian	T 3
P 148	Propotamochoerus sp.	Suidae	Siwalik Group, Middle Division	Pliocene	S 51b
P 149	Prosophinctes	Ammonite	Kanthkot Sandstone, Kutch	Jurassic	K 23
P 150	Proteacidites cf. P. miniporatus	PF	Inter-trappean beds of Anjar, Gujarat	Cretaceous	i 8- 2a
P 151	Proteacidites reticulatus				
P 152	Protobuntonia hartmani	Ostracoda	Inter-trappean beds of Rajahmundry, Andhra Pradesh	Cretaceous	i 8- 1b
P 153	Protocardia pauli	Lamellibranchia	Wadhwan Formation	Cretaceous	W 1
P 154	Protocytherella sp.	Crustacea	Khalsi Formation, J & K	Cretaceous	K 42
P 155	Protohertzina anabarica	TF	Tethyan Basin, Himalaya	Cambrian	T 20
P 156	Protolepidodendron sp.	Pteridophyta (Clubmoss)	Muth Formation, Kashmir, Kishtwar, Ladakh and Kumaon	Devonian	M 66
P 157	Protoretepora ampla	Bryozoa	Po Formation, Spiti-Kinnaur, Himachal Pradesh	Carbonniferous	P 38
			Umaria Marine Bed and Manendragarh Marine Bed	Permian	U 5 & M22
			Seti Khola Formation, Bhutan		S 25
P 158	Protragocerus	Bovidae	Siwalik Group, Middle Division	Pliocene	S 51b
P 159	Proxapertites cursus	PF	Inter-trappean beds of Padwar/Ranipur, Madhya Pradesh	Cretaceous	i 8- 4c
P 160	Proxapertites operculatus				

Fossil No.	Fossil Name	Fossil Group	Formation and area	Formation Age	Formation Number
P 161	Psammichnites	TF	Kunzun La Formation, Spiti	Lr. Cambrian	K 73a
P 162	Pseudofusilina	Foraminifera	Khunamuh Formation, J &K	Permian	K 51
P 163	Pseudoharrisichara cf. P. baytikshanesis	Charophyta (Algae)	Inter-trappean beds of Mamoni, Rajasthan	Cretaceous	i 8- 6a
P 164	Pseudomaura nobilis	Gastropoda	Anaipadi Formation, Trichinopoly Gr, Trichinopoly	Cretaceous	A 22
P 165	Pseudomonotis	Lamellibranchia	Giumal Formation, Himachal Pradesh	Cretaceous	G 15
P 166	Pseudomonotis griesbachi		Ophiceras Zone, Spiti	Triassic	O 6
P 167	Pseudomonotis himaica		Hedenstroemia Beds, Spiti		H 12
			Rambakot Member, Uttarakhand		R 13
P 168	Pseudopolygnatus primus	Conodonta	Tethyan Basin, Himalaya	Devonian	T 20
P 169	Pseudopoly-morphina devarapalleensis	Foraminifera	Infra-trappean (=Lameta) beds of Marepalli, Andhra Pradesh	Cretaceous	I 5-1a
P 170	Pseudotextularia cf. elegans	Foraminifera	Chikkim Formation, Himachal Pradesh	Cretaceous	C 31
P 171	Pseudotextularia elegans		Infra-trappean beds, Rajahmundry		i 5-1b
P 172	Psiloceras	Ammonite	Lamayuru Formation, Ladakh, J&K	Lowermost Jurassic	L 8
P 173	Psilophyton	Plantae (Tracheo-phyta - a vesicular plant)	Takche Formation, Spiti	Ordovician-Silurian	T 3
P 174	Psilospora	Insecta	Umia Formation, Kutch	Cretaceous	U 8
P 175	Psygmophyllum haydeni	Ginkogales	Gangamopteris Beds, Gondwana, Kashmir	Permian	G 5
			Nishatbagh Bed, J & K		N 25
P 176	Pterinidium sp. Cf. P. simplex	Pteridophyta	Krol Group, Nainital Syncline	Neoproterozoic	K 64
P 177	Ptilophyllum	Pteridophyta	Raghvapuram Formation	Lr. Cretaceous	R 2
			Vemavaram Formation	Cretaceous	V 7
			Kota Formation	Jurassic	K 62

Fossil No.	Fossil Name	Fossil Group	Formation and area	Formation Age	Formation Number
			Gangapur Formation.	Lr. Cretaceous	G 6
			Umia Formation, Kutch, Gujarat	Cretaceous	U 8
P 178	Ptilophyllum acutifolium		Terani Plant Beds, Up. Gondwana, Trichinopoly	Lr. Cretaceous	T 14
			Dubrajpur Formation	Triassic	D 51
			Raghvapuram Formation	Lr. Cretaceous	R 2
			Tirupati Formation	Lr. Cretaceous	T 34
			Pariwar Formation, jaisalmer, Rajasthan	Cretaceous	P 23
P 179	Ptilophyllum cutchense		Raghvapuram Formation	Lr. Cretaceous	R 2
P 180	Ptilophyllum tennerrimum		Inter-trappean beds, Rajmahal Basin	Cretaceous	i 8-9
P 181	Ptychites aff. Everesti	Cephalopoda	Gushal Formation, Himachal Pradesh	Triassic	G 35
P 182	Ptychites gerardi		Daonella Shales, Spiti		D 9
P 183	Ptychites rugifer		Upper Muschekalk, Spiti		U 16
			Hoti Member, Uttarakhand		H 20
P 184	Ptychites scheibleri		Rambakot Member, Uttarakhand		R 13
P 185	Ptychoparia	Trilobita	Kunzun La Formation, Zanskar, J&K	Cambrian	K 73b
			Kunzun La Formation, Spit	Mid. Cambrian	K 73a
P 186	Ptychophylloceras	Ammonite	Umia ammonite bed, Kutch	Cretaceous	U 6
			Lower Katrol, Kutch	Jurassic	L 29
P 187	Pycnoditidae gen. et sp. indet	Fish	Inter-trappean beds of Asifabad, Rajahmundry and Naskal, Andhra Pradesh; Kisalpuri, Madhya Pradesh; Nagpur, Maharashtra	Cretaceous	i 8- 1a,b,c; 4a; 5b
P 188	Pycnodonta bouchevoni	Lamellibranchia	Wadhwan Formation, Gujarat	Cretaceous	W 1
P 189	Pyctonoporidium	Algae	Kallakkudi Formation	Cretaceous	K 11

Fossil No.	Fossil Name	Fossil Group	Formation and area	Formation Age	Formation Number
Q 1	Quadrum qartineri	NF	Karai Shale Formation and Maravattur Clay Formation	Cretaceous	K 26 M 28
Q 2	Quinqueloculina	Foraminifera	Niniyur Formation, Trichinopoly	Paleocne	N 23
			Quaternary sediments of Gujarat Mainland	Quaternary	Q 5
			Kankawati Series, Kutch	Pliocene	K 21
			Infra-trappean (=Lameta) beds of Rajahmundry, Andhra Pradesh	Cretaceous	i 5-1b
Q 3	Quinqueloculina semilunum		Miliolite / Porbandar Stone, Gujarat	Pleistocene	M 49
Q 4	Qyilonipollenites	PF	Cuddalore Sandstone Formation, Neyveli Lignite Field, Tamil Nadu	Miocene	C 56
R 1	Racemiguembelina fructicosa	Foraminifera	Infra-trappean beds, Rajahmundry	Cretaceous	i 5-1b
R 2	Radiastrea sp.	Coral	Tethyan Basin, Himalaya	Silurian	T 20
R 3	Raja sudhakari	Fish	Inter-trappean beds of Asifabad, Andhra Pradesh	Cretaceous	i 8- 1a
R 4	Rajahia mamalensis	Plantae	Mamal Formation, J & K	Lr. Permian	M 19
R 5	Ramapithecus	Primates	Siwalik Group, Middle Division	Pliocene	S 51b
R 6	Rana pusilla	Amphibian (Frog)	Inter-trappean beds, Mumbai	Cretaceous	I 8-5a
R 7	Raniganjia bengalensis	Pteridophyta	Gondwana Supergroup, Lower Division	Permian to Triassic	G 23a
R 8	Ranoidea indet	Amphibian	Inter-trappean beds of Naskal, Andhra Pradesh	Cretaceous	i 8- 1c
R 9	Raoellid Artiodactyls	Vertebrata	Subathu Group, Katakot, Metka, Moghala inliers, Rajauri District, J & K	Eocene	S 65
R 10	Raphistoma	Gastropoda	Thango Formation, Himachal Pradsh. Ordovician	Ordovician	T 23

Fossil No.	Fossil Name	Fossil Group	Formation and area	Formation Age	Formation Number
R 11	Rastellum (Arcostrea) pectinatum	Lamellibranchia	Kallankurichi Formation, Ariyalur Gr, Trichinopoly	Cretaceous	K 13
R 12	Raymondella	NA	Ordovician-Silurian rocks of Shans Abani Syncline, Handwar, J & K	Ordovician	O7
R 13	Rechnisaurus	Reptilia	Yerrapalli Formation	Triassic	Y 5
R 14	Redlichia noelingi	Trilobita	Kunzun La Formation, Spit	Mid. Cambrian	K 73a
			Tal Formation, Himachal Pradesh	Lr. Cambrian	T 5
R 15	Refinesquina	Brachiopoda	Ordovician-silurian rocks of Shans Abani Syncline, Handwar, J & K	Ordovician	O 7
			Shiala Formation, Uttarakhand		S 31
R 16	Reineckeia	Cephalopoda	Anceps beds, Kutch	Jurassic	A 23
			Rehmanni beds, Kutch		R 31
R 17	Resserella	Brachiopoda	Ordovician-Silurian rocks of Shans Abani Syncline, Handwar, J & K	Ordovician	O 7
R 18	Reticularia	Brachiopoda	Agglomeratic Slate, J & K	Carboniferous-Permian	A 4
			Umaria Marine Bed and Manendragarh Marine Bed, Madhya Pradesh	Permian	U 5 & M22
R 19	Reticularia lineate		Po Formation, Spiti-Kinnaur, Himachal Pradesh	Carbonniferou s	P 38
R 20	Rhabdochara sp.	Algae	Inter-trappean beds of Biligi, Karnataka	Cretaceous	i 8- 3a
R 21	Rhacopteris ovate	Pteridosper-matophyta	Po Formation, Spiti-Kinnaur, Himachal Pradesh	Carbonniferous	P 38
R 22	Rhinoceros	Rhinocerotidae (Mammalia)	Quaternary sediments of Andhra Pradesh, flood plain of river Pampa	Pleistocene	Q 3
			Siwalik Group, Upper Division		S 51c

Fossil No.	Fossil Name	Fossil Group	Formation and area	Formation Age	Formation Number
R 23	Rhinoceros unicornis		Pleistocene sediments of Kurnool, Ghod, Manjra, Pravara and Godavari		P 37a
R 24	Rhinoceros palaeindicus		Pleistocene sediments of Lower Karewa, Pinjor of Upper Siwaliks		P 37d
R 25	Rhizamospora	Plantae (Angiospermae)	Barakar Formation, Pranhita-Godavari Valley	Lr. Permian	B 24
R 26	Rhizocorallium	TF	Po Formation, Spiti-Kinnaur area	Carboniferous	P 38
R 27	Rhizomys	Rodentia	Siwalik Group, Lower, Middle & Upper Divisions	Miocene, Pliocene & Pleistocene	S51 a,b,c
R 28	Rhodophyta	Algae	Bagh Beds (Nimar Sandstone) Narmada Basin	Cretaceous	B 8
R 29	Rhombodus sp.	Fish	Inter-trappean beds of Asifabad and Naskal, Andhra Pradesh	Cretaceous	i 8- 1a,c
R 30	Rhombopora	Bryozoa	Umaria Marine Bed and Manendragarh Marine Bed	Permian	U 5, M 22
R 31	Rhynchonella (Wilsonia) cf. omega	Brachiopoda	Muth Formation, Byans, Uttarakhand	Devonian	M 66
R 32	Rhynchonella bambanagensis		Coral Limestone, Spiti	Triassic	C 47
R 33	Rhynchonella griebachi		Basal muschelkalk, Spiti		B 31
			Niti Limestone Member, Uttarakhand		N 26
R 34	Rhynchonella sp.		Gushal Formation, Himachal Pradesh		G 35
R 35	Robulus cf. occidentalis	Foraminifera	Inter-trappean beds of Rajahmundry, Andhra Pradesh	Cretaceous	i 8- 1b
R 36	Robulus sp. Indet				
R 37	Rotalia	Foraminifera	Broach Formation, Gujarat	Pliocene	B 72
			Quaternary sediments of Gujarat Mainland	Quaternary	Q 5
			Aida Stage, Kutch	Oligocene	A 6

Fossil No.	Fossil Name	Fossil Group	Formation and area	Formation Age	Formation Number
R 38	Rotaliapora appenninica	Foraminifera	Chikkim Formation, Himachal Pradesh	Cretaceous	C 31
R 39	Rotalipora cushmani		Uttatur Gr. Trichinopoly		U 21
			Karai Shale Formation and Maravattur Clay Formation		K 26 M 28
R 40	Rouaultia	TF	Thango Formation, Kinnaur,Himachal Pradesh	Ordovician	T 23
R 41	Rsophycus	TF	Kunzun La Formation, Spiti	Lr. Cambrian	K 73a
R 42	Rugoglobigerina hexacamerata	Foraminifera	Kallankurichi Formation, Trichinopoly	Cretaceous	K 13
R 43	Rugoglobigerina rugosa		Baratang Group, Andaman and Nicobar Islands		B 25
			Infra-trappean beds, Rajahmundry		i 5-1b
R 44	Rusophycus	TF	Thango Formation, Kinnaur,Himachal Pradsh	Ordovician	T 23
			Kunzun La Formation, Spiti	Lr. Cambrian	K 73a
			Garbyang Formation, Uttarakhand	Cambrian	G 8
R 45	Rutihesperipites trochuensis	PF	Inter-trappean beds of Padwar/Ranipur, Madhya Pradesh	Cretaceous	i 8- 4c
R 46	Ruzophycus	TF	Po Formation, Spiti-Kinnaur area	Carboniferous	P 38
S 1	Sahnites	PF	Lapthal Member, Uttarakhand	Jurassic	L 13
S 2	Sahnitherium rangapuri	Mammalia	Inter-trappean beds of Rangapur, Andhra Pradesh	Cretaceous	i 8- 1d
S 3	Salmoniformorum rectangulus	Fish	Inter-trappean beds of Naskal and Rangapur, Andhra Pradesh	Cretaceous	i 8- 1c,d
S 4	Samaropsis	Cordaitales	Gangamopteris Beds, Gondwana, Kashmir	Permian	G 5
			Nishatbagh Bed, J & K		N 25

Fossil No.	Fossil Name	Fossil Group	Formation and area	Formation Age	Formation Number
			Bahreli Formation, Arunachal Pradesh		B 10
S 5	Sarasinella subspinosa	Cephalopoda	Habur Formation, Jaisalmer, Rajasthan	Cretaceous	H 1
S 6	Saurichthys	Fish	Kalhal Formation, Himachal Pradesh	Triassic	K 9
S 7	Sauropod eggshell fragments	Dinosauria	Inter-trappean beds of Asifabad, Andhra Pradesh; Anjar, Gujarat and Nagpur, Maharashtra	Cretaceous	i 8- 1a; 2a; 5b
S 8	Scapanorhynchus baghensis	Fish (Shark teeth)	Bagh Beds (Amba Dongar) Narmada Basin	Cretaceous	B 8
S 9	Scaphites kingeanus	Ammonite	Karai Shale Formation and Maravattur Clay Formation	Cretaceous	K 26 M 28
S 10	Schellwienella willigmi	Brachiopoda	Muth Formation, Malla Johar, Uttarakhand	Devonian	M 66
S 11	Scheuringi-pollenites	PF	Barakar Formation, Pranhita-Godavari Valley	Lr. Permian	B 24
S 12	Schizoneura gondwanensis	Pteridophyta (Equisetales)	Gangamopteris Beds, Gondwana, Kashmir	Permian	G 5
			Gondwana Supergroup, Lower Division	Permian to Triassic	G 23a
			Bahreli Formation, Arunachal Pradesh	Permian	B 10
			Barakar Formation, Pranhita-Godavari Valley	Lr. Permian	B 24
S 13	Schloenbachia iNAlate	Ammonite	Karai Shale Formation and Maravattur Clay Formation	Cretaceous	K 26 M 28
S 14	Schwagerina	Foraminifera	Khunamuh Formation, J &K	Permian	K 51
S 15	Scleropages	Fish	Takli Formation, Nagpur	Cretaceous-Paleocene	T 4
S 16	Serpulids	Invertebrata (Worms, an Annelida)	Karai Shale Formation and Maravattur Clay Formation & Kallakkudi Limestone Formation Uttatur Gr, Trichinopoly	Cretaceous	K 26 M 28 K 11

Fossil No.	Fossil Name	Fossil Group	Formation and area	Formation Age	Formation Number
S 17	Serranidarum sp.	Fish	Inter-trappean beds of Rangapur, Andhra Pradesh; Lakshmipur, Gujarat; Nagpur, Maharashtra	Cretaceous	i 8- 1d; 2d; 5b
S 18	Shoreoxylon	Plantae (Fossil wood)	Baripada Beds, Orissa	Miocene	B 26
S 19	Shweboemys (Carteremys) leithii	Reptilia	Inter-trappean beds of Bombay, Maharashtra	Cretaceous	i 8- 5a
S 20	Sibirites prahlada	Cephalopoda	Lower Muscelkalk, Spiti	Triassic	L 31
			Niti Member, Uttarakhand		N 27
S 21	Siderolites	Foraminifera	Mahdek Formation, Assam/Meghalaya	Cretaceous	M 4
S 22	Siderolites calcitrpoides	Foraminifera	Kallankurichi Formation, Ariyalur Group, Trichinopoly	Cretaceous	K 13
S 23	Siluriformes indet	Fish	Inter-trappean beds of Naskal, Andhra Pradesh; Kisalpuri, Madhya Pradesh	Cretaceous	i 8- 1c; 4a
S 24	Sivajiceras	Cephalopoda	Rehmanni beds, Kutch	Jurassic	R 31
			Pachcham coral bed, Kutch		P 2
S 25	Sivanasua	Carnivora	Siwalik Group, Middle Division	Pliocene	S 51b
S 26	Sivapithecus	Primates	Siwalik Group, Lower Division	Miocene	S 51a
S 27	Sivatherium	Giraffidae	Siwalik Group, Upper Division	Pleistocene	S 51c
S 28	Sivatherium giganteum		Pleistocene sediments of Lower Karewa, Pinjor of Upper Siwaliks		P 37d
S 29	Skilolithos Haldermann	TF	Bhuj Beds, Kutch, Gujarat	Cretaceous	B 53
S 30	Skolithos	TF	Thango Formation Kinnaur, Himachal Pradesh	Ordovician	T 23
			Kunzun La Formation, Spiti	Lr. Cambrian	K 73a
			Po Formation, Spiti-Kinnaur area	Carboniferous	P 38

Fossil No.	Fossil Name	Fossil Group	Formation and area	Formation Age	Formation Number
S 31	Solenoporacea	Algae	Kallakkudi Limestone Formation, Uttatur Gr, Trichinopoly	Cretaceous	K 11
S 32	Sorites	Foraminifera	Khari Series, Kutch	Miocene	K 44
S 33	Sowerbyella	Brachiopoda	Ordovician-Silurian rocks of Shans Abani Syncline, Handwar, J & K	Ordovician	O 7
			Shiala Formation, Uttarakhand		
S 34	Sphaerulites indica	Lamellibranchia (Rudistt)	Kallakkudi Limestone Formation, Uttatur Gr, Trichinopoly	Cretaceous	K 11
S 35	Sphenoceramus	Lamellibranchia	Bagh Beds, Narmada Basin	Cretaceous	B 8
S 36	Sphenophyllum	Pteridophyta	Gangamopteris Beds, Lower Gondwana, Kashmir	Permian	G 5
S 37	Sphenopteridium fucillatum	Pteridospermatophyta	Po Formation, Spiti-Kinnaur, Himachal Pradesh	Carbonniferous	P 38
S 38	Sphenopteris hislopi	Fern	Gondwana Supergroup, Upper Division	Mid. Triassic-Lr. Cretaceous	G 23b
S 39	Sphenopteris polymorpha		Gangamopteris Beds, Lower Gondwana, Kashmir	Permian	G 5
S 40	Spheroidinella sp.	Foraminifera	Inter-trappean beds of Rajahmundry, Andhra Pradesh	Cretaceous	i 8- 1b
S 41	Spinizonocolpites echinatus	PF	Inter-trappean beds of Mohgaon-Kalan, Madhya Pradesh; Lalitpur, Uttar Pradesh	Cretaceous	i 8- 4b; 7a
S 42	Spinizonocolpites baculatus		Inter-trappean beds of Padwar/Ranipur, Madhya Pradesh		i 8- 4c
S 43	Spirellus shankeri	Invertebrata (Arthopoda)	Tethyan Basin, Himalaya	Cambrian	T 20
S 44	Spirifer	Brachiopoda	Fenestella Shale Formation, Kashmir	Carboniferous	F 2
			Saser Brangsa Formation	Carboniferous	S 13

Fossil No.	Fossil Name	Fossil Group	Formation and area	Formation Age	Formation Number
			Boulder Slate Formation, Garhwal, Uttarakhand	Permian	B 71
			Yamne Formation, Arunachal Pradesh		Y 2
			Badhaura Formation, Bikaner-Nagaur Basin, Rajasthan		B 5
			Umaria Marine Bed and Manendragarh Marine Bed		U 5 & M22
			Gushal Formation, Himachal Pradesh	Triassic	G 35
S 45	Spiriferella rajah	Brachiopoda	Salooni Formation, Lahaul, Himachal Pradesh	Permian	S 4
			Gungri Formation, Himachal Pradesh		G 33
			Boulder Slate Formation, Garhwal, Uttarakhand		B 71
			Kuling Formation, Uttarakhand	Permian	K 69
S 46	Spiriferina	Brachiopoda	Boulder Slate Formation, Garhwal, Uttarakhand	Permian	B 71
S 47	Spiriferina	Brachiopoda	Umaria Marine Bed and Manendragarh Marine Bed	Permian	U 5, M 22
S 48	Spiriferina griesbachi		Monotis Shales, Spiti	Triassic	M 57
			Coral Limestone, Spiti		C 47
			Kioto Member, Uttarakhand		K 55
S 49	Spiriferina shalshalensis		Grey Beds, Spiti		G 28
S 50	Spiriferina stracheyi		Lower Muscelkalk, Spiti		L 31
			Hoti Member, Uttarakhand		H 20
S 51	Spirigera hunica	Brachiopoda	Rimkhim Member, Uttarakhand	Triassic	R 36
S 52	Spirigera maniensis		Quartzite Series, Spiti		Q 2

Fossil No.	Fossil Name	Fossil Group	Formation and area	Formation Age	Formation Number
S 53	Spiroclypous ranjanae	Foraminifera	Waior Stage, Kutch	Oligocene	W 2
S 54	Spiroloculina	Foraminifera	Quaternary sediments of Gujarat Mainland	Quaternary	Q 5
			Miliolite / Porbandar Stone, Gujarat	Pleistocene	M 49
S 55	Spiroplectamminia	Foraminifera	Sriperumbudur Formation, Tamil Nadu	Mid. Jurassic (?)	S 63
S 56	Spiticeras aff. bulliformis	Ammonite	Spiti Formation	Jurassic	S 61
S 57	Spiticeras aff. eximus				
S 58	Spiticeras cf. subspitiensis				
S 59	Spiticeras eximius				
S 60	Spiticeras guttatus				
S 61	Spiticeras indicus				
S 62	Spiticeras sp.		Giumal Formation	Cretaceous	G 15
S 63	Spiticeras spitiensis		Spiti Formation	Jurassic	S 61
S 64	Spondylus latus	Lamellibranchia	Wadhwan Formation, Gujarat	Cretaceous	W 1
S 65	Stegodon	Proboscidea	Siwalik Group, Upper Division	Pleistocene	S 51c
S 66	Stegodon insignis		Quaternary sediments of Tamil Nadu, Soyamali	Pleistocene	Q 9
S 67	Stegodon insignis ganesa		Quaternary sediments of central India	Lr. To Up. Pleistocene	Q 4
			Pleistocene sediments of Central Narmada (upper group), Palmar	Pleistocene	P 37b
S 68	Stegolophodon	Equidae	Siwalik Group, Lower Division	Miocene	S 51a
S 69	Stegosaurus	Dinosauria	Trichinopoly Group, Trichinopoly	Cretaceous	T 44
S 70	Stellatopollis araripensis	PF	Inter-trappean beds of Padwar/Ranipur, Madhya Pradesh	Cretaceous	I 8- 4c
S 71	Stephanochara cf. S. levis	Charophyta (Algae)	Inter trappean beds of Kora, Gujarat; Gurmatkal, Karnataka	Cretaceous	i 8- 2c; 3b

Fossil No.	Fossil Name	Fossil Group	Formation and area	Formation Age	Formation Number
S 72	Stigmatoygus elatus	Echinoidea	Kallankurichi Formation, Trichinopoly	Cretaceous	K 13
S 73	Streblites	Ammonite	Lower Katrol, Kutch Gajansar Beds, Kutch	Jurassic	L 29 G 2
S 74	Streblites aff. plicodiscus		Spiti Formation		S 61
S 75	Streblopteris	Invertebrata (Crinoidea-Phylum Echinodermata)	Umaria Marine Bed and Manendragarh Marine Bed	Permian	U 5 & M22
S 76	Streblus beccari	Foraminifera	Miliolite / Porbandar Stone, Gujarat	Pleistocene	M 49
S 77	Streptelasmids	Coral	Takche Formation, Spiti	Silurian-Ordovician	T 3
S 78	Striatites	PF	Lapthal Member, Uttarakhand	Jurassic	L 13
S 79	Striatopodo-carpites	PF	Spiti Member, Uttarakhand	Up. Jurassic	S 62
S 80	Strophalosia	Brachiopoda	Boulder Slate Formation, Garhwal, Uttarakhand	Permian	B 71
S 81	Strophomena	Brachiopoda	Ordovician-Silurian rocks of Shans Abani Syncline, Handwar, J & K	Ordovician	O 7
S 82	Stygmatopygus elatus	Echinoidea (Phylum echinodermata)	Ottakovil Formation, Ariyalur Group, Trichinopoly	Cretaceous	O 9
S 83	Stylina	Actinozoa	Pachcham coral bed, Kutch	Jurassic	P 2
S 84	Subansiria naganensis	Brachiopoda	Miri Formation, Arunachal Pradesh	Permian	M 54
S 85	Subdichoto-moceras sp.	Ammonite	Spiti Formation	Jurassic	S 61
S 86	Subtiliolithus kachchhensis	Dinosauria	Inter-trappean beds of Kisalpuri, Madhya Pradesh	Cretaceous	i 8- 4a
S 87	Sudamericidae indet	Mammalia	Inter-trappean beds of Naskal, Andhra Pradesh	Cretaceous	i 8- 1c
S 88	Sueroceras bomtense	Cephalopoda	Miri Formation, Arunachal Pradesh	Permian	M 54
S 89	Sugrivapithecus	Primates	Siwalik Group, Lower Division	Miocene	S 51a

Fossil No.	Fossil Name	Fossil Group	Formation and area	Formation Age	Formation Number
S 90	Sunetta scripta	Lamellibranchia	Chilka Lake deposit of estuarine shells, Orissa	pleistocene	C 33
S 91	Sus	Suidae	Siwalik Group, Upper Division	Pleistocene	S 51c
S 92	Sus namadicus		Pleistocene sediments of Central Narmada (Lower group)	Pleistocene	P 37b
S 93	Sus namadicus		Conglomerate Bed, Narmada valley	Middle Pleistocene	C 46
S 94	Sus namadicus		Quaternary sediments of central India	Mid. Pleistocene	Q 4
S 95	Sus palaeindicus			Up. pleistocene	
S 96	Sus sps. (Body parts)		Quaternary sediments of Andhra Pradesh, Gunglakamma and Munneru river basins	Pleistocene	Q 3
S 97	Syringothyris cuspidata	Brachiopoda	Tethyan Basin, Himalaya	Carboniferous	T 20
			Syringothyris Lst. Formation, Kashmir	Carboniferous	S 79
			Lipak Formation	Carboniferous	L 24
S 98	Syringothyris lydekkeri		Lipak Formation	Carboniferous	L 24
T 1	Taeniolthasrus permixtus	NA	Agglomeratic Slate Formation, Kashmir	Carboniferous-Permian	A 4
T 2	Taeniopteris danaeoides	Cycadophyta	Gondwana Supergroup, Lower Division	Permian to Triassic	G 23a
T 3	Taeniopteris feddeni		Gangamopteris Beds, Gondwana, Kashmir	Permian	G 5
T 4	Taeniopteris kashmirensis		Gangamopteris Beds, Gondwana, Kashmir	Permian	G 5
T 5	Taeniopteris sp.	Cycadophyta	Intertrappeans, Dubrajpur Formation, Rajmahal Basin	Triassic	i 8-9
T 6	Taeniopteris spatulata		Raghvapuram Formation	Lr. Cretaceous	R 2
			Sivaganga Beds, Tamil Nadu	Lr. Cretaceous	S 49
			Inter-trappean beds, Rajmahal Traps	Cretaceous	i 8-9
T 7	Taeniopteris vittata		Pariwar Formation, jaisalmer, Rajasthan	Cretaceous	P 23

Fossil No.	Fossil Name	Fossil Group	Formation and area	Formation Age	Formation Number
T 8	Talicypridea	Crustacea	Inter-trappean beds of Asifabad, Nagpur, Mamoni and Gurmatkal	Cretaceous	i 8- 1a; 5b; 6a; 3b
T 9	Taphrosphys sp.	Reptilia	Inter-trappean beds of Kisalpuri, Madhya Pradesh	Cretaceous	i 8- 4a
T 10	Taramelliceras	Ammonite	Lower Katrol, Kutch	Jurassic	L 29
T 11	Taramelliceras sp.	Ammonite	Spiti Formation	Jurassic	S 61
T 12	Teichichnus	TF	Thango Formation Kinnaur,Himachal Pradesh	Ordovician	T 23
T 13	Tellina	Lamelli-branchia	Coastal Gondwana Formations, Ramnathpuram, Tamil Nadu	Lr.Cretaceous	C 45
T 14	Tellina cf. strigata	Lamelli-branchia	Giumal Formation, Himachachal Pradesh	Cretaceous	G 15
T 15	Telmatodon	Anthracoth-eriidae	Siwalik Group, Lower Division	Miocene	S 51a
T 16	Tentaculites	Gastropoda	Lipak Formation	Carboniferous	L 24
T 17	Terebratulids	Brachiopoda	Trichinopoly Gr. Trichinopoly	Cretaceous	T 44
			Kulakkalnattam Formation, Trichinopoly		K 69
T 18	Tetraconodon	Suidae	Siwalik Group, Middle Division	Pliocene	S 51b
T 19	Tetragramma	Echinoidea	Bagh Beds, Narmada Basin	Cretaceous	B 8
T 20	Textularia	Foraminifera	Qazil Langer Formation, Jammu and Kashmir	Cretaceous	Q 1
			Chikkim Formation, Himachal Pradesh		C 31
T 21	Textularia cf. conica		Miliolite / Porbandar Stone, Gujarat	Pleistocene	M 49
T 22	Thalassinoide	TF	Subathu Formation, Uttarakhand/ Himachal Pradesh	Eocene to Miocene	S 65
			Ottakovil Formation, Ariyalur Group, Trichinopoly	Cretaceous	O 9
			Trichinopoly Group, Trichinopoly	Cretaceous	T 44

Fossil No.	Fossil Name	Fossil Group	Formation and area	Formation Age	Formation Number
			Dagshai Formation, Himachal Pradesh	Eocene-Oligocene	D 1
T 23	Thalassinoides Ehrenburg		Bhuj Beds, Kutch, Gujarat	Cretaceous	B 53
T 24	Thalmannita madrugaensis	Foraminifera	Niniyur Group/ Formation, Trichinopoly	Early Paleocene	N 23
T 25	Thamnastrea	Actinozoa	Pachcham coral bed, Kutch	Jurassic	P 2
T 26	Thecosmilia clathrata	Cephalopoda	Tagling Limestone, Spiti	Jurassic	T 1
T 27	Theropod eggshell fragments	Dinosauria	Inter-trappean beds of Anjar, Gujarat; Kisalpuri, Madhya Pradesh	Cretaceous	i 8- 2a; 4a
T 28	Thinnfeldia	Pteridospermatophyta	Inter-trappean beds, Rajmahal Basin	Cretaceous	i 8-9
T 29	Thinnfeldia khatangiensis		Gondwana Supergroup, Upper Division	Mid. Triassic-Lr. Cretaceous	G 23b
T 30	Tibetella	Foraminifera	Hundri Formation, Jammu and Kashmir	Cretaceous	H 21
T 31	Tiggilites	TF	Subathu Formation, Himachal Pradesh	Eocene-Miocene	S 65
T 32	Tilanosaurus	Reptilia	Kallamedu Formation, Trichinopoly	Cretaceous	K 12
T 33	Timiriaservia	Ostracoda	Inter-trappean beds, Mumbai	Cretaceous	i 8-5a
T 34	Titanosaurus	Reptilia	Ariyalur Group, Trichinopoly	Cretaceous	A 35
T 35	Todisporites kutchensis	PF	Inter-trappean beds of Lalitpur, Uttar Pradesh	Cretaceous	I 8- 7a
T 36	Todisporites major				
T 37	Toeniocrada sp.	Plantae	Muth Formation, Kashmir, Kishtwar, Ladakh and Kumaon	Devonian	M 66
T 38	Tolypammina	Foraminifera	Umaria Marine Bed and Manendragarh Marine Bed	Permian	U 5 & M22
T 39	Topaeum	Ammonite	Ukra Beds, Umia Formation, Kutch	Cretaceous	U 2
T 40	Toroisporis major	PF	Inter-trappean beds of Padwar/Ranipur, Madhya Pradesh	Cretaceous	i 8- 4c

Fossil No.	Fossil Name	Fossil Group	Formation and area	Formation Age	Formation Number
T 41	Torquantisp-hinctes	Ammonite	Kanthkot Sandstone, Kutchl, Katrol Formation, Kutch	Jurassic	K 23
			Spiti Member, Uttarakhand		S 62
			Middle Katrol, Kutch		M 45
T 42	Torquatisphinctes inetermedius		Spiti Formation		S 61
T 43	Torquatisphinctes torquatus				
T 44	Trachycardia sp.	Lamellibranchia	Subathu Formation, Uttarakhand	Eocene to Miocene	S 65
T 45	Tragocerus	Bovidae	Siwalik Group, Upper Division	Pleistocene	S 51c
T 46	Tramelliceras	NA	Dhosa Oolite, Chari Formation, Kutch	Jurassic	D 34
T 47	Trapezoidella sp.	Crustacea	Infra-trappean (=Lameta) beds of Pisdura, Maharashtra	Cretaceous	i 5-4b
T 48	Traumatocrinus	Invertebrata (Crinoidea)	Rimkhim Member, Uttarakhand	Triassic	R 36
T 49	Triangulipoll turonicusis	PF	Inter-trappean beds of Padwar/Ranipur, Madhya Pradesh	Cretaceous	i 8- 4c
T 50	Tricollaraeporites	PF	Cuddalore Sandstone Formation, Neyveli Lignite Field, Tamil Nadu	Miocene	C 56
T 51	Trigonia	Lamellibranchia	Trigonia Beds, Kutch	Cretaceous	T 45
			Kanthkot Sandstone, Kutch, Katrol Formation, Kutch	Jurassic	K 23
			Pachcham shell Lst, Kutch		P 4
			Pachcham Basal Beds, Kutch		P 1
			Umia Stage, Kutch, Gujarat	Cretaceous	U 11
			Kulakkalnattam Formation, Trichino-poly Gr, Trichinopoly		K 69
T 52	Trigonotreta	Brachiopoda	Umaria Marine Bed and Manendragarh Marine Bed	Permian	U 5 & M 22

Fossil No.	Fossil Name	Fossil Group	Formation and area	Formation Age	Formation Number
			Seti Khola Formation, Bhutan		S 25
T 53	Triloculina	Foraminifera	Kankawati Series, Kutch	Pliocene	K 21
			Niniyur Formation, Trichinopoly	Paleocne	N 23
			Quaternary sediments of Gujarat Mainland	Quaternary	Q 5
T 54	Triloculina cf. laevigata		Inter-trappean beds of Rajahmundry, Andhra Pradesh	Cretaceous	I 8- 1b
T 55	Triloculina trigonula		Niniyur Group/ Formation, Trichinopoly	Early Paleocene	N 23
T 56	Trilophodon	Proboscidea	Siwalik Group, Upper Division	Pleistocene	S 51c
T 57	Triporoletes reticulates	PF	Inter-trappean beds of Asifabad, Nagpur, Mamoni & Gurmatkal	Cretaceous	i 8- 1a; 5b; 6a; 3b
			Inter-trappean beds of Padwar/Ranipur, Madhya Pradesh		i 8- 4c
T 58	Trizygia speciosa	Pteridophyta	Mammal Formation, J & K	Lr. Permian	M 19
T 59	Trochammina	Foraminifera	Umaria Marine Bed and Mahendragarh Marine Bed	Permian	U 5, M 22
			Surma Group, Assam and adjoining areas	Early Miocene	S 73
T 60	Trochonems	Gastropoda	Thango Formation Kinnaur, Himachal Pradesh. Ordovician	Ordovician	T 23
T 61	Tropites subbulatus	Gastropoda	Tropites Beds, Spiti	Triassic	T 47
T 62	Trumcorotaloides rohre	Foraminifera	Babia Stage, Kutch	Eocene	B 3
T 63	Turborotalia centralis	Foraminifera	Babia Stage, Kutch	Eocene	B 3
T 64	Turrilites costatus	Ammonite	Uttatur Gr. Trichinopoly	Cretaceous	U 21
			Karai Shale Formation and Maravattur Clay Formation, Trichinopoly		K 26 M 28

Fossil No.	Fossil Name	Fossil Group	Formation and area	Formation Age	Formation Number
T 65	Turritella	Gastropoda	Tirupati Formation	Lr. Cretaceous	T 34
			Kand Formation, Gujarat Mainland	Lr. Miocene	K 19
T 66	Tylostoma globosum	Gastropoda	Bagh Beds, Narmada Basin	Cretaceous	B 8
U 1	Uhligites aff. adunta	Ammonite	Spiti Formation	Jurassic	S 61
U 2	Umiatites	Ammonite	Umia ammonite bed, Kutch	Cretaceous	U 6
U 3	Unio deccanensis	Lamellibranchia	Inter-trappean beds of Biligi, Gurmatkal and Upparhatti, Karnataka; Mandla, Madhya Pradesh	Cretaceous	i 8- 3a,b,c; 4d
			Infra-trappean (=Lameta) beds of Dongargaon, Maharashtra		i 5-4a
U 4	Uraloceras siangense	Cephalopoda	Miri Formation, Arunachal Pradesh		M 54
U 5	Uvigerina	Foraminifera	Khari Series, Kutch	Miocene	K 44
V 1	Vaginulina cf. V. cenomana	Foraminifera	Infra-trappean (=Lameta) beds of Rajahmundry, Andhra Pradesh	Cretaceous	i 5-1b
V 2	Venericardia cf. mutabilis	Lamellibranchia	Subathu Formation, Uttarakhand	Eocene to Miocene	S 65
V 3	Ventilabrella glabrata	Foraminifera	Baratang Group, Andaman and Nicobar Islands	Cretaceous	B 25
V 4	Vermiporella	Algae	Takche Formation, Spiti	Silurian-Ordovician	T 3
V 5	Vertebraria indica	Pteridospermatophyta	Gangamopteris Beds, Gondwana, Kashmir	Permian	G 5
			Bahreli Formation, Arunachal Pradesh		B 10
			Barakar Formation, Pranhita-Godavari Valley		B 24
V 6	Vesicaspora	PF	Gondwana Supergroup, Lower Division	Permian to Triassic	G 23a
V 7	Virgatosphinctes	Ammonite	Upper Katrol Sandstone, Kutch	Jurassic	U 14

Fossil No.	Fossil Name	Fossil Group	Formation and area	Formation Age	Formation Number
V 8	Virgatosphinctes	Ammonite	Umia ammonite bed, Kutch	Cretaceous	U 6
V 9	Virgatosphinctes communis	Ammonite	Spiti Formation	Jurassic	S 61
V 10	Virgatosphinctes densiplicatus				
V 11	Virkkipollenites sp.	PF	Gondwana Supergroup, Lower Division	Permian to Triassic	G 23a
V 12	Vishnutherium	Giraffidae (Mammalia)	Siwalik Group, Middle Division	Pliocene	S 51b
V 13	Vitreisporites	PF	Katrol Formation, Kutch	Jurassic	K 38
V 14	Viverra chinjiensis	Carnivora	Siwalik Group, Lower Division	Miocene	S 51a
V 15	Viviparus sp.	Gastropoda	Inter-trappean beds of Gurmatkal and Upparhatti, Karnataka; Sagar, Madhya Pradesh	Cretaceous	i 8- 3b,c; 4e
V 16	Voysa aff. Scalaroideum	Gastropoda	Anaipadi Formation, Trichinopoly Gr, Trichinopoly	Cretaceous	A 22
W 1	Waaganoconcha	Brachiopoda	Permian rocks, Niti, Malla Johar and Byan areas, Uttarakhand	Permian	P 31
W 2	Waaganoconcha cf. humboldti		Seti Khola Formation, Bhutan	Permian	S 25
W 3	Waaganoconcha cf. vagens				
W 4	Waaganophyllum	Coral	Gechang Formation, Himachal Pradesh	Permian	G 10
W 5	Waagenophyllum indicum		Lahaul, Himachal Pradesh		K 67
W 6	Waagenia	Cephalopoda	Lower Katrol & Middle Katrol, Kutch	Jurassic	L 29 M 45
W 7	Wadiasaurus	Reptilia	Yerrapalli Formation	Triassic	Y 5
W 8	Warthia cf. intermedia	Gastropoda	Miri Formation, Arunachal Pradesh	Permian	M 54
W 9	Weichselia reticulate	Fern	Dharangadhara Group, Gujarat	Cretaceous	D 28
W 10	Wentzellella cf. salinaria	Coral	Kukti Formation, Lahaul, Himachal Pradesh	Permian	K 68
W 11	Williamsonia	Cycadophyta	Umia Beds, Kutch, Gujarat	Cretaceous	U 7

Fossil No.	Fossil Name	Fossil Group	Formation and area	Formation Age	Formation Number
			Vemavaram Formation		V 7
W 12	Wyndhamia	Brachiopoda	Agglomeratic Slate, J & K	Carboniferous-Permian	A 4
X 1	Xaniognathus nenwpossensis	Mammalia	Triassic of NW Himalaya, Kashmir, Ladakh, Spiti	Up. Triassic	T 40
X 2	Xenaspis	Cephalopoda	Productus Shale, Spiti	Permian	P 51
X 3	Xenaspis carbonaria		Gungri Formation, Himachal Pradesh		G 33
X 4	Xenodiscus nivalis	Cephalopoda	Hedenstroemia Beds, Spiti	Triassic	H 12
X 5	Xestoleberis rupnarayan-alurensis	Ostracoda	Niniyur Formation	Paleocne	N 23
X 6	Xestoleberis sp.		Inter-trappean beds of Rajahmundry, Andhra Pradesh	Cretaceous	i 8-1b
Y 1	Yulghoensis	NA	Muth Formation, Spiti	Devonian	M 66
Z 1	Zaphrentis beyrichi	Actinozoa	Tethyan Basin, Himalaya	Permian	T 20
Z 2	Zonocypris spirula	Ostracoda	Inter-trappean beds of Kora and Lakshmipur, Gujarat; Gurmatkal, Karnataka	Cretaceous	i 8-2c,d; 3b
Z 3	Zoophycos	TF	Gungri Formation, Himachal Pradesh	Permian	G 33

■■■■

VI

VI LIST OF FOSSILIFEROUS FORMATIONS WITH THEIR FOSSILS

(Total number of Fossiliferous formations in this list are 177)

S.N.	Formation No.	Formation Name: Fossil Numbers
1	A 4	Agglomeratic Slate: A 90; B 36, 41, 44; C 124; D 19, 23; E 37; F 10; M 24; P 130; R 18; T 1; W 12
2	A 6	Aida Stage: C 43; M 45; R 37
3	A 22	Anaipadi Formation: C 17; D 2; E 29; K 21; P 1, 164; V 16
4	A 23	Anceps Beds: H 50; i 14; K 15; P 69; R 16
5	A 35	Ariyalur Group: A 11; E 28, 43; G 30, 68; i 16; K 3; L 43; T 34
6	A 42	Athleta Beds: O 20; P 60
7	B 3	Babia Stage: G 35, 71, 78; T 62, 63
8	B 5	Badhaura Formation: B 10; C 70; D 34; P 124; S 44
9	B 8	Bagh Beds: A 28, 68; B 16; C 23, 39, 63, 87, 102; G 17; i 16; L 5; M 61; N 1; P 47, 93, 131, 144; R 25; S 8, 35; T 19, 60
10	B 10	Bahreli Formation: C 78; G 10, 48, 53, 55; P 83; S 4, 12; V 5
11	B 14	Baisakhi Formation: K 9; L 44, 45, 46
12	B 24	Barakar Formation: C 88; D 21; F 1; G 5, 48; i 15; R 25; S 11, 12; V 5
13	B 25	Baratang Group: G 40; N 42; R 43; V 3
14	B 26	Baripada Beds: C 94; P 16; S 18
15	B 31	Basal Muschelkalk: R 33
16	B 42	Bhadasar Formation: K 9; L 44, 45, 46
17	B 53	Bhuj Beds: P 16, 98; S 29; T 23
18	B 58	Bijori Formation: G 48, 64
19	B 71	Boulder Slate Formation: A 89; C 110; D 37, 38, 50; F 5; M 10; N 22; P 127, 136;p S 44, 45, 46, 80
20	B 72	Broach (Bharuch) Formation: B 23; C 43; E 12; N 36; r 37

S.N.	Formation No.	Formation Name: Fossil Numbers
21	C 16	Chari Formation: A 12, 47; C 2, 69; G 24; P 120
22	C 26	Chhojan Member: B 5; P 37
23	C 27	Chhongtash Formation: H 43, 47; S 93
24	C 31	Chikkim Formation: D 22; G 38, 42, 45; H 13; i 16; L 3, 11; N 30; P 18, 170; R 38; T 20
25	C 33	Chilka Lake: A 50; C 99; M 30; P128; S 90
26	C 45	Coastal Gondwana Formation: A 87; i 16; O 36; P 51, 52; T 13
27	C 46	Conglomerate Bed, Narmada Valley: H 43, 47; S 93
28	C 47	Coral Limestone: R 32; S 48
29	C 56	Cuddalore Sandstone Formation: C 24, 96; D 43; P 119; Q 4; T 50
30	D 1	Dagshai Formation: C 41; T 22
31	D 8	Daonella Limestone: D 8
32	D 9	Daonella Shales: D 9; P 182
33	D 28	Dharangadhara Group: C 99; M 15; O 6; W 9
34	D 34	Dhosa Oolite: D 47; E 17; M 16; P 38, 59; 71; T 46
35	D 40	Dinod Formation: D 44; N 43
36	D 48	Dras Formation: H 44; O 14, 41
37	D 49	Dras Volcanics: A 19; D 29; G 37; H 24
38	D 50	Drumgam Formation: B 8
39	D 51	Dubrajpur Formation: G 23; P 178
40	F 2	Fenestella Shale: F 4; P 56, 136; S 14
41	G 1	Gaj Formation: M 47; N 26; P 84
42	G 2	Gajansar Beds: A 52; B 7; H 7, 31; S 73
43	G 5	Gangamopteris Beds: A 7, 21; C 36, 77, 79; G 7, 13, 44, 51, 53, 63; K 5, 6, 11; L 18; P 13, 55, 83, 140, 175; S 12, 36, 39; T 3, 4; V5
44	G 6	Gangapur Formation: E 4; G 22; P 8, 177
45	G 8	Garbyang Formation: C 95; G 79; i 18; P 100, 118; R 44; S 4
46	G 10	Gechang Formation: D 18; E 37, 39, 41; N 21; W 4
47	G 15	Giumal Formation: B 12; C 20; G 69; K 14; N 9, 11; O 2, 36; P 56, 165; S 62; T 14
48	G 22	Gondwana of Extrapeninsula: G 6, 48
49	G 23	Gondwana Supergroup: A 8; B 32, 42; C 52, 54, 74; D 21; F 1; G 8, 11, 53; N 33; P 64; R 7; S 12, 38; T 2, 29; V 6, 11
50	G 28	Grey Beds: J 2; S 49
51	G 31	Gund Formation: A 51, 55; L 15, 17
52	G 33	Gungri Formation: C 110; M 11; P 137; S 45; X 3; Z 3

S.N.	Formation No.	Formation Name: Fossil Numbers
53	G 35	Gushal Formation: J 1; K 16; P 181; R 34; S 44
54	H 1	Habur Formation: D 24; S 5
55	H 4	Halobia Beds: H 3; J 3
56	H 12	Hedenstroemia Beds: F 7; H 14; P 167; X 4
57	H 20	Hoti Member: K 13; P 183; S 50
58	H 21	Hundri Formation: A 15; M 32; O 14; P 19, 129; T 30
59	i 5	Infra-trappean Beds: A 1, 17, 20, 22; C 11, 13, 15, 45, 75, 101, 108, 112, 113, 115, 119, 126, 129, 133; D 11, 30; E 32; F 8; G 39, 44, 46, 74; L 54; L 54, 56; M 29, 40, 51 to 54; N 6; P 15, 17, 20 to 23, 37, 42, 75, 87, 88, 89, 91, 103, 104,, 122, 169, 171; Q 2; R 1, 43; T 47; U 3; V 1
60	**i** 8	Inter-trappean Beds: A 16, 18, 29, 31, 40, 42 to 46, 60, 73, 78,, 94; B 1, 2, 33, 45; C 9, 10, 12, 14, 22, 31, 32, 34, 55, 67, 85, 91, 98,, 101, 103, 106, 107, 111, 113, 114, 116, 117, 118, 120 to 123, 128, 129, 130; D 5, 6, 7, 10, 12, 14, 15, 45; E 1, 19, 31, 33; F 8; G 1, 2, 3, 33, 58 to 61, 66, 67, 83; H 9, 10, 19, 22, 23, 48, 49; i 5, 7, 11, 13; L 1, 2, 4, 20, 23, 30, 38, 55; M 3, 7, 8,, 27, 41, 49, 50; N 6, 31, 35, 37, 39; O 7, 19, 21, 34, 40; P 8, 11, 12, 14, 24, 37, 40, 41, 42, 54, 65 to 68, 72, 77, 90, 92, 105 to 109,, 117, 150, 151, 152,, 159, 160,, 163, 180, 187; R 3, 6, 8, 20, 29, 35, 36,, 45; S 2, 3, 7, 17, 19, 23, 40, 41, 42,, 70, 71, 86, 87; T 5, 6, 8, 9, 27, 28, 33, 35, 36, 40, 49,, 54, 57; U 3; V 15; X 6; Z 2
61	J 22	Juvavites Beds: J 4
62	K 5	Kakdi stage: G 26, 27, 34, 36
63	K 9	Kalhal formation: S 6
64	K 11	Kallakkudi Formation: A 4, C 6, 46; E 27, 34; H 12, 46; L 10, 37; M 43; O 42; P 56, 189; S 16, 31, 34
65	K 12	Kallamedu Formation: B 39; M 28; T 32
66	K 13:	Kallankurichi Formation: E 43; G 41; H 11; L 16; P 3, 82; R 11, 42; S 22, 72
67	K 19	Kand/Kond Formation: : A 34; B 9; C 18, 72; L 13; M 47; O 36; P 56; T 65
68	K 21	Kankawati Series: A 25; Q 2; T 53
69	K 23	Kanthkot Sandstone: A 71; B 15; C 81; E 17; P 141, 149; T 41, 51
70	K 26	Karai Shale Formation: A 24; B 6; C 4, 3790; E 30; H 17; L 9; M 6, 13, 42; P 56, 102; Q 1; R 39; S 9, 13, 16; T 64
71	K 38	Katrol Formation: A 12, 47; C 2, 69; G 24; P 120; V 13
72	K 42	Khalsi Formation: C 25, 127; D 1; L 12,, 49; O 15 to 18; P 154
73	K 44	Khari series: A 25, 56, 88; D 46; G 77; L 14; S 32; U 5
74	K 51	Khunamuh Formation: C 53, 109; N 16; P 43, 126, 162; S 14
75	K 55	Kioto Member: A 30; H 4; P 134; S 48
76	K 62	Kota Formation: G 48; K 22; P 177

S.N.	Formation No.	Formation Name: Fossil Numbers
77	K 64	Krol Formation: B 11; i 1; P 176; W 5
78	K 68	Kukti Formation: C 110; M 36; W 10
79	K 69	Kulakkalnattam Formation: A 13; B 35; C 89, 97, 125; G 82; K 20; L 24; M 33; N 3; P 44, 86, 94; S 45; T 17, 51
80	K 70	Kuling Formation: C 3; P 116
81	K 73	Kunzun La Formation: A 9, 32, 33; C 95; D 4, 33, 39, 40, 42; E 3, 13; F 9; H 51; L 29, 33, 58, M 55; N 17; O 5, 31; P 57, 78, 95, 99, 161, 185; R 14, 41, 44; S 30
82	L 8	Lamayuru Formation, Ladakh, J & K: P 172
83	L 13	Lapthal Member: D 51; M 1; P 116; S 1, 78
84	L 24	Lipak Formation: A 76; C 104; i 10; N 24; O 29; P 74, 135, 136; S 97, 98; T 16
85	L 29	Lower Katrol: H 55; P 186; S 73; T 10; W 6
86	L 31	Lower Muschelkalk: K 13; S 20, 50
87	M 2	Macrocephalus Beds: B 4; D 51; i 12; K 2; M 2; P 115
88	M 4	Mahadek Formation: G 37, 72; O 13; S 21
89	M 19	Mamal Formation: G 19, 52, 54, 57; L 41; P 46; R 4; T 58
90	M 22	Manendragarh Marine Bed: A 72, 91, 92; D 18; E 35, 36; G 47; H 57; P 114, 136, 157; R 18, 30; S 44, 47, 75; T 38, 52, 59
91	M 28	Maravattur Clay Formation: A 11; B 6; C 4, 37, 90; E 30; H 17; L 9; M 6, 42; P 56, 102; Q 1; R 39; S 9, 13, 16; T 64
92	M 39	Meekoceras Zone: M 20, 22
93	M 40	Megalodon Limestone/Para Stage: D 25; M 25
94	M 45	Middle Katrol: A 63; K 18; P 25; T 41, W 6
95	M 49	Miliolite/Porbandar Stone: B 23; c 44; E12; G 28, 29; L 48; N 34; Q 3; S 54,, 76; T 21
96	M 54	Miri Formation: A 37, 89; B 10; C 71; D 19, 35; E 37; F 4; L 32, 35; O 3; P 73, 110, 113, 125; S 84, 88; U 4; W 8
97	M 57	Monotis Shales: M 56; S 48
98	M 66	Muth Quartzite/Formation: A 77; B 18; C 8, 56, 57, 104; F 2; H 37; i 4; N 18; O 24, 26, 27, 30; P 63, 135, 156; R 31; S 10; T 37; Y 1
99	N 4	Nagri Formation: B 37
100	N 23	Niniyur Group/Formation: A 5, 54; B 3; C 43, 60; D 13; G 32; L 57; N 4, 27; P 39; Q 2; T 24, 53, 55; X 5
101	N 25	Nishatbagh Bed: C 80; G 9, 14, 55, 56; P 175; S 4
102	N 26	Niti Limestone Member: R 33
103	N 27	Niti Member: S 20

S.N.	Formation No.	Formation Name: Fossil Numbers
104	O 6	Ophiceras Zone: O 11; P 166
105	O 7	Ordovician-Silurian rocks: A 38; L 7, 8, 21, 22; O 22, R 12, 15, 17; S 33, 81
106	O 8	Otoceras Zone: O 38, 39
107	O 9	Ottakovil Formation: D 55; G 39, 43, 73; O 12; P 2; S 82; T 22
108	P 1	Pachcham Basal Beds: C 76; E 14; T 51
109	P 2	Pachcham Coral Bed: M 2, 57; P 133; S 24, 83; T 25
110	P 4	Pachcham Shell Limestone: C 76; M 2; T 51
111	P 5	Pachmarhi Formation: K 15; M 35; P 49, 76
112	P 21	Parahio Series: P 7, 70
113	P 23	Pariwar Formation: E 5; M 31; O 37; P 178; T 7
114	P 31	Permian rocks: L 31; W 1
115	P 36	Piram Beds: D 17
116	P 37	Pleistocene sediments: A 57; B 26, 27, 40; C 16, 28; E 2, 8, 9, 10, 22, 25; H 27, 28; P 9; R 23, 24; S 28, 67, 92
117	P 38	Po Formation: A 69, 85; C 59; D 34; F 4; G 80; P 81, 101, 113, 138, 139, 157; R 19, 21, 26, 46; S 30, 37
118	P 51	Productus Shale: C 109; M 11; X 2
119	Q 1	Qazil Langer Formation: G 31; T 20
120	Q 2	Quartzite Series: S 62
121	Q 3	Quaternary sediments of Andhra Pradesh: B 28; c 26, 92; E 26; R 22; S 96
122	Q 4	Quaternary sediments of Central India: A 39, 93; B 29; C 27, 29; E 9, 10, 11, 21, 23; H 29, 30; S 67, 94, 95
123	Q 5	Quaternary sediments of Gujarat Mainland: A 24; E 12; N 30, 36; Q 2; R 37; S 54; T 53
124	Q 8	Quaternary sediments of Maharashtra: E 7, 24; H 42
125	Q 9	Quaternary sediments of Tamil Nadu: B 27; E 9; L 34; S 66
126	R 2	Raghvapuram Formation: A 23, 49; C 2, 48, 66, 69; H 8; i 6; L 52; M 39; O 37; P 177, 178, 179; T 6
127	R 9	Rajmahal Traps: C 50; D 31
128	R 12	Ramania Stage: H 26; O 9, 10
129	R 13	Rambakot Member: E 18; F 7; M 21, 22; O 11, 39; P 167, 184
130	R 23	Raniganj Formation: G 49, 50
131	R31	Rehmanni Beds: i 9; K 12, R 16; S 24
132	R 36	Rimkhim Member: D 8; H 3; S 51; T 48
133	S 4	Salooni Formation: M 48; P 136; S 45

S.N.	Formation No.	Formation Name: Fossil Numbers
134	S 5	Sancha Malla Group: A 2; N 9, 10
135	S 13	Saser Brangsa/Shaksgam Formation: A 74; C 84; F 4; L 31; M 11; S 44
136	S 25	Seti Khola Formation: N 25; P 111, 157; T 52; W 2, 3
137	S 31	Shiala Formation: B 10; C 5; O 22, 29; R 15
138	S 37	Sillakudi Formation: B 24; C 21, 89; G 30; i 8; K 4, 7; P 61
139	S 49	Sivaganga Beds: B 31; C 51; E6; G 18; P 121; T 6
140	S 50	Sivaganga Formation: i 8
141	S 51	Siwalik Group: A 6, 10, 27, 35, 36;; B 17, 25, 34; C 7, 26, 38, 40, 61, 68, 73, 82, 93; D 16, 49, 52; F 3; G 4, 15, 16, 20, 62; H 21, 39, 40, 41, 53, 60; K 1, 23; L 36, 39, 50, 53; N 28; P 27, 48, 143, 148, 158; R 5, 22, 27; S 25, 26, 27, 65, 68, 89, 91; T 15, 18, 45, 56;; V 12, 14
142	S 61	Spiti Formation/Shale: A 2, 64, 65, 80, 82, 83, 84; B 8, 20, 21, 22; D 48; E 17; G 67, 75, 76; H 32, 36, 45, 52; i 17; K 10, 17, 18, 19; L 25, 26; M 17; N 9, 10; P 5,6, 28 to 36,, 142; S 56 to 61, 63, 74, 85; T 11, 42, 43; U 1; V 9, 10
143	S 62	Spiti Member: A 81; B 19; C 83, 100; H 33, 35; P 116; S 79; T 41
144	S 63	Sriperumbudur Formation: H 8; P 53; S 55
145	S 65	Subathu Formation/Group: A 66, 67; B 38; C 30, 41; i 19; N 41, 41; O 8, 36; P 10, 58, 96; R 9; T 22, 31, 44; V 2
146	S 73	Surma Group: H 8; M 44
147	S 79	Syringothyris Limestone: B 14; C 42, 56; H 38; N 18; O 43; P 123, 136; S 97
148	T 1	Tagling Limestone: M 26; T 26
149	T 3	Takche/Manchap Formation: A 26, 62, 74; C 42, 62, 105; F 2, 4; G 21; H 2, 5, 15; O 22; P 26, 146, 147, 173; S 77; V 4
150	T 4	Takli Formation: A 14, 61; H 25; L 19; M 19; N 7, 38; O 35; P 106; S 15
151	T 5	Tal Group/Formation: C 64; L 27; O 1; R 14
152	T 6	Talchir Formation: D 34; E 36, 38, 40, 42; F 6; G 48; P 132
153	T 14	Terani Plant Bed: E 5; P 178
154	T 18	Tertiary of Gujarat: D 27
155	T 20	Tethyan Basin: A 41, 70, 75; C 1, 56 to 59, 110; D 3, 41; E 37; H 6, 16, 56; i 2, 3; L 6, 28, 42; M 4, 48; N 18; O 4, 25, 28, 32, 33; P 79, 80, 112, 155, 168; R 2; S 43, 97; Z 1
156	T 23	Thango Formation: A 59; B 23; M 55; P 62, 97, 145, 147; R 10, 40, 44; S 30; T 12, 60
157	T 34	Tirupati Formation: C 9, 33; D 26, 31; E 4; G 23; L 51; M 9; N 3, 29; P 121, 178; T 65
158	T 40	Triassic of NW Himalaya: M 33; N 12 to 15, 19, 20; X 1

S.N.	Formation No.	Formation Name: Fossil Numbers
159	T 44	Trichinopoly Group: A 58; G 81; P 85; S 69; T 17,22
160	T 45	Trigonia Beds: T 51
161	T 47	Tropites Beds: D 36; T 61
162	U 2	Ukra Beds: A 86; C 35, 65; T 39
163	U 5	Umaria Marine Bed: A 72, 91, 92; D 18; E 35, 36; G 47; H 57; P 114, 136, 157; R 18, 30; S 44, 47, 75; T 38, 52, 59
164	U 6	Umia ammonite Bed: H 20; M 37; P 186; U 2; V 8
165	U 7	Umia Beds: B 30; C 47; E 4, 20; G 22; M 5; O 7; W 11
166	U 8	Umia Formation: A 48; C 2; D 20; i 6; L 52; M 38,, 59; P 120, 174, 177
167	U 11	Umia Stage: A 3, 79; G 25; H 58, 59; L 40, T 51
168	U 14	Upper Katrol Sandstone: A 81; V 7
169	U 15	Upper Katrol Shales: D 53; H 34
170	U 21	Uttatur Group: H 1; M 12; O 42; P 56, 183; R 39; T 64
171	V 5	Variegated Formation: F 2; O 23
172	V 7	Vemavaram Formation: D 7, 32; i 16; M 18; N 3; P 177; W 11
173	W 1	Wadhwan Formation: A 53; C 86; P 153, 188; S 64
174	W2	Waior Stage: O 9; P 45; S 53
175	Y 2	Yamne Formation: A 89; B 43; E 15, 16; G 65; L 47; M 23, 58, 60; N 5, 8, 23; P 136; S 44
176	Y 5	Yerrapalli Formation: P 50; R 13; W 7
177	Z 4	Zewan Formation: C 109; N 16

■■■■

VII

LIST OF FOSSIL GROUPS PERTAINING TO THIS BOOK

S.N	Fossil Group	Number of fossils
1	Actinozoa or Anthozoa	7
2	Algae	24
3	Ammonites	125
4	Amphibia	8
5	Anthracotheriidae	10
6	Artiodactyla	8
7	Bovidae	8
8	Brachiopoda	82
9	Bryozoa	12
10	Carnivora	11
11	Cephalopoda	66
12	Charophyta	13
13	Coniferales or Pinales	12
14	Conodonta	27
15	Coral	14
16	Cordaitales	7
17	Crocodylidae	5
18	Crustacea	5
19	Cycadophyta	8
20	Deleted (fossil deleted due to lack of conformity)	1
21	Dinosaur	9
22	Echinoidea	6
23	Equidae	9
24	Ferns	20

S.N	Fossil Group	Number of fossils
25	Fishes	39
26	Foraminifera	137
27	Fungi	3
28	Gastropoda	53
29	Ginkgoales	2
30	Giraffidae	7
31	Gymnospermae	4
32	Insecta	5
33	Invertebrata	11
34	Labyrinthodontia	4
35	Lamellibranchia or Pelicypoda or Bivalvia	79
36	Mammalia	8
37	NA (fossils for which fossil group could not be found by the author)	15
38	NF (Nanofossil)	3
39	Ostracoda	70
40	Palm or Palmae or Arecacea	4
41	PF (Palynofossil)	65
42	Plantae or plant fossil	28
43	Porifera	3
44	Primates	8
45	Proboscidea	14
46	Pteridophyta	14
47	Pteridospermatophyta	26
48	Reptilia	15
49	Rhinocerotidae	9
50	Rodentia	5
51	Suidae	13
52	TF (Tracefossil)	49
53	Trilobita	19
54	Vertebrata	8
	TOTAL	1217

■■■■

VIII

DETAILS OF FOSSIL GROUPS

(Total number of Fossil Groups in this list are 54)

(Pertaining to this book only)

All the living organisms have been classified into following 5 kingdoms:

1. Kingdom Monera: These organisms donot have a defined nucleus or organelles, nor do any of them show multicellular body designs and have prokaryotic cell organization. In this kingdom are incuded bacteria, blue-green algae or cyanobacteria and micoplasma.
2. Kingdom Protista: This kingdom include many kinds of unicellular eukarytic organisms, such as algae, diatoms and protozoans.
3. Kingdom Fungi: This kingdom includes heterotrophic eukaryotic organisms, such as yeast and mushrooms.
4. Kingdom Plantae: These are multicellular eukaryotes with cell walls, and perform photosynthesis using chlorophyll and include all plants.
5. Kingdom Animalia: These are multicellular eukaryotes without cell walls and are heterotrophs and include all the animals.

Below are given details of only those fossil groups under which fossils mentioned in this book have been placed.

S.N.	The Fossil Group	Brief description
1	Actinozoa	Belongs to phylum Coelentrata of the invertebrate animals and includes sea anemones, corals, etc.
2	Algae	Belongs to the phylum Thallophyta of the Plant Kingdom and forms a large diverse group ranging from unicellular to multicellular forms.
3	Ammonites	Ammonites are an extinct group of subclass Ammonoidea
4	Amphibia	Forms a class of the vertebrate animals and includes Labyrinthodontia which is an extinct subclass of class Amphibia
5	Anthracotheriidae	It is an extinct family of hippopotamus like animals which belonged to Artiodactyla-Ungulata of class Mammalia.
6	Artiodactyla	These are even-toed ungulates of class Mammalia of the vertebrate animals.

S.N.	The Fossil Group	Brief description
7	Bovidae	It is a family of order Artiodactyla, class Mammalia, and includes animals like bison, antelopes, buffalo, domestic cattle, goats, sheep, etc.
8	Brachiopoda	These are invertebrate bivalve animals, which have an upper and a lower hard shell also called a valve.
9	Bryozoa	These are aquatic invertebrate animals (commonly moss animals) and form a phylum having scientific name as Ectoprocta.
10	Carnivora	Carnivora forms an order of class Mammalia of Vertebrata and are essentially meat-eating animals.
11	Cephalopoda	These are marine animals with bilateral body symmetry and form a class of phylum Mollusca of the invertebrate animals.
12	Charophyta	This forms a division of fresh water algae
13	Coniferales or Pinales	These belong to class Gymnospermae of the Plant Kingdom and are distinguished by presence of cone or related structure for reproduction.
14	Conodonta	Conodonta is an extinct class of primitive vertebrates
15	Coral	Coral is a marine invertebrate animal of class Actinozoa or Anthozoa of phylum Coelentrata. Corals typically live in compact colonies formed by individual 'polyps' and give rise to well known coral reefs.
16	Cordaitales	Cordaitales is an order of class Gymnospermae, these plants are extinct now and were dominant during the Carboniferous period.
17	Crocodylidae	Crocodylidae forms an order of class Reptilia of the vertebrate animals.
18	Crustacea	Crustacea is a class of phylum Arthopoda of the invertebrate animals.
19	Cycadophyta	This is a sub-class of class Gymnospermae, these are ever-green plants with large hard and stiff leaves and stout-woody trunk.
20	Deleted	Fossil details deleted due to lack of conformity
21	Dinosauria or Dinosaur	Belongs to class Reptilia of phylum Chordata. These animals lived during Jurassic and Cetaceous periods and were characterised by large eyes and long narrow skull. Carnosauria and Coelurosauria are also Dinosaurs
22	Echinoidea	A Class of phylum Echinodermata of the invertebrate animals.
23	Equidae	Equidae is a family of order Perissodactyla of class Mammalia and includes horse and related animals.
24	Ferns	Ferns or Filicales have xylem, stem, leaves, roots and belongs to the class Pteropsida of phylum Pteridophyta of the Plant Kingdom.
25	Fishes	Fishes or Pisces forms a class of phylum Chordata of the Animal Kingdom.

S.N.	The Fossil Group	Brief description
26	Foraminifera	Forms an order of unicellular invertebrate aquatic animals of phylum Protozoa having thin pseudopodia for catching food and also have an external shell.
27	Fungi	Is itself a Kingdom, different from plant and animal
28	Gastropoda	Gastropoda forms a class of phylum Mollusca of the invertebrate animals and includes snails and slugs which are single shelled aquatic animals.
29	Ginkgoales	Ginkgoales forms an order of subclass Coniferophyta of class Gymnospermae of the Plant Kingdom.
30	Giraffidae	Giraffidae is a family of ruminant artiodactyls mammals
31	Gymnospermae	Group of seed producing plants, having naked seeds, not enclosed in an ovary. Gymnospermae forms a class of phylum spermatophyte and includes Conifers, Cycades, Ginkgo and Gnetales.
32	Insecta	It is a Class of phylum Arthopoda of the Invertebrate animals, charectarised by having chitinous exoskeleton, a three part body (head, thorax, and abdomen), three pairs of jointed legs, compound eyes, and one pair of antennae.
33	Invertebrata	A sub-kingdom of kingdom animalia. Invertebrate includes animals without backbones or the vertebral column.
34	Labyreinthodontia	Labyrinthodontia is an extinct amphibian subclass
35	Lamellibranchia or Pelecypoda or Bivalvia	A class of phylum Mollusca. These are marine or fresh water invertebrate animals with laterally compressed bodies enclosed by a shell in two hinged parts.
36	Mammalia	Mammalia is a class of vertebrate animals, which are warm blooded and are distinguished by presence of hair, three middle ear bones, mammary glands in females and a neocortex (a region of brain).
37	NA	Fossil Group could not be found by the author
38	NF	Nanofossil
39	Ostracoda	Belongs to class Crustacea of phylum Arthopoda. These are small bivalve crustaceans.
40	Palm or Palmae or Arecaceae	A family of flowering plants belonging to the Monocot Angiosperms, which includes perennial lianas, shrubs and trees, commonly known as palms.
41	PF	Palynofossil
42	Plantae	Plant fossil for which further taxonomy could not be found by the author
43	Porifera	Porifera is one of the phylum of the invertebrate animals. These are multicellular organisms with bodies full of pores and channels for water circulation.

S.N.	The Fossil Group	Brief description
44	Primates	Primates form an order of the class Mammalia of the Animal Kingdom. Among Primates are included placental mammals, such as apes, monkeys, chimpanzees, etc. and also the Homo sapiens i.e. the human beings.
45	Proboscidea	Proboscidea is an order of class Mammalia of the vertebrate animals and includes only one living family Elephantidae (the elephant family) and several extinct families. Presence of trunk and tusks is generally a characterstic feature of these animals.
46	Pteridophyta	Pteridophyta is a phylum of sub-kingdom Cryptogamae of kingdom Plantae. In Pteridophyta plants, true leaves and roots are present and spores borne on the undersurface of the leaf e.g. ferns. Lycopodiales, equisetales, sphenophyllales, ferns / filicales, etc. are included in the phylum pteridophyta.
47	Pteridospermatophyta	Pteridospermatophyta forms a sub-class of class Gymnospermae. These plants lived during Paleozoic and Mesozoic era and are extinct now.
48	Reptilia	Reptilia is a class of vertebrate animals, which originated during the Carboniferous period, these are cold-blooded and egg-laying animals which includes dinosaurs, crocodiles, tortoise, etc., etc.
49	Rhinocerotidae	Constitute a family of class Mammalia and order Perissodactyla
50	Rodentia	Rodents belong to order Rodentia of class Mammalia of the vertebrate animals and include such small animals like rats, squirrels, etc. Rodents are characterised by two constantly growing incisors in each of the upper and lower jaws which must be kept short by gnawing.
51	Suidae	Suidae is a family of order Artiodactyla of class Mammalia of the vertebrate animals. The Suidae includes domestic pigs as well as wild pigs. The earliest Suidae fossils are found in the Oligocene formations.
52	TF	Trace Fossil
53	Trilobita	A class of phylum Arthopoda. The Trilobites (means three lobes) were marine animals (extinct now) which lived during Paleozoic (Cambrian to Permian).
54	Vertebrata	A sub-kingdom of kingdom animalia. Vertebrata includes animals having backbones or the vertebral column.

IX

LIST OF FOSSILS AS PER THE FOSSIL GROUPS

(This list cosists of 54 Fossil Groups which include 1217 fossil genera)

1. Actinozoa : M 57; P 145, 146, 147; S 83; T 25; Z 1 =7
2. Algae : A 4, 5, 20, 53, 54, 62; C 32, 39, 64, 102, 105; D 13; E 27; G 21, 66, 83; H 1, 46; P 39, 189; R 20, 28; S 31; V 4 = 24
3. Ammonite : A 2, 3, 13, 63, 64, 65, 79 to 84, 86, 87; B 4 to 8, 12, 19 to 22; C 4, 63, 65, 85; D 2, 24, 47, 48; E 17, 28, 29, 30; G 73, 75, 76; H 11, 20, 31 to 36, 45, 52; K 3, 4, 8, 9, 10, 14, 17, 18, 19, 20, 21; L 24, 40; M 6, 16, 17, 33, 37; N 1, 2, 9, 10, 11; O 2; P 1 to 6, 28 to 36, 47, 51, 52, 53, 69, 93, 94, 144, 149, 172, 186; S 9, 13, 56 to 63, 73, 74, 85; T 10, 11, 39, 41, 42, 43, 65; U 1, 2; V 7, 8, 9, 10 = 125
4. Amphibian : A 40, 45; i 11; L 23; P 50, 92; R 6, 8 = 8
5. Anthracotheriidae : A 35, 36; C 26, 27, 28, 29, 40; H 21, 41; T 15 = 10
6. Artiodactyla : A 39; B 40; H 27, 28, 29, 30, 42, 43 = 8
7. Bovidae : B 17, 25, 26, 27; D 52; G 15; P 158; T 45 = 8
8. Brachiopoda : A 72, 74, 75, 76, 77; B 36, 44; C 21, 42, 89; D 23, 34, 35, 36, 37, 38; L 6, 7, 8, 21, 22, 27, 28, 29, 31, 32, 35; M 10, 11; N 8, 21, 22, 23, 24, 25; O 1, 3, 22, 23, 24, 25, 26, 29, 30; P 44, 62, 63, 135, 136, 137, 138, 139; R 15, 17, 18, 19, 31, 32, 33, 34; S 10, 33, 44, 45, 46, 47, 48, 49, 50, 51, 52, 80, 81, 84, 97, 98; T 17, 52; W 1, 2, 3, 12 = 82
9. Bryozoa : D 50; F 4, 5, 6; H 2; L 51; P 124, 125, 126, 127, 157; R 30 =12
10. Carnivora : A 10, 25, 27; B 28, 29; C 16, 93; F 3; L 53; S 25; V 14 =11
11. Cephalopoda : A 71; B 15; C 35, 109, 110; D 51, 53; E 18; F 7; G 67; H 4, 7, 14, 50; I 9, 12, 14; J 1, 2, 3, 4; K 2, 12, 13, 15, 16; M 1, 2, 20, 21, 22, 25; N 3, 4; O 11, 20, 27, 28, 38, 39, 41; P 38, 59, 60, 71, 84, 115, 133, 134, 141, 142, 181, 182, 183, 184; R 16; S 5, 20, 24, 88; T 26; U 4; W 6; X 2, 3, 4 = 66
12. Charophyta : H 9; M 40, 41; N 6, 7; P 54, 105, 106, 107, 108, 109, 163; S 71 =13
13. Coniferales : A 48, 49; B 30, 31, 32, 33, 42; E 4, 5, 6; P 8, 119 = 12
14. Conodonta : A 9, 26; B 14, 18; C 56, 57, 58; F 9; H 37, 38; I 2, 3; L 42; M 34; N 12, 13, 14, 15, 16, 18, 19, 20; O 5, 43; P 26, 123, 168 = 27
15. Coral: C 1, 62, 104; E 34; F 2; H 5, 6, 15; M 43; R 2; S 77; W 4, 5, 10 = 14

16. Cordaitales : C 77, 78, 79; K 11; N 32, 33; S 4 = 7
17. Crocodylidae : A 14, 61; C 91, 92; P 9 = 5
18. Crustacea : C 33, 106; P 154; T 8, 47 = 5
19. Cycadophyta : O 37; T 2, 3, 4, 5, 6, 7; W 11 = 8
20. Deleted (fossil details deleted due to lack of conformity) : D 54 = 1
21. Dinosauria : B 39; K 22; M 19, 27; O 21; S 7, 69, 86; T 27 = 9
22. Echinoidea : C 46; H 18; P 61; S 72, 82 : T 19 = 6
23. Equidae : C 82; E 21, 22, 23, 24, 25, 26; H 39; S 68 = 9
24. Fern : C 47, 48, 49, 50, 51, 52; D 30; G 22, 23, 24, 25; L 15, 18, 52; M 9; O 7; P 55; S 38, 39; W 9 = 20
25. Fish : A 42, 43, 44, 60; C 9, 31, 53, 55; D 10; G 17; H 10, 48; i 5; L 2, 5, 19, 20; N 37, 38; O 34, 35; P 11, 12, 13, 65, 66, 67, 68, 72, 131, 187; R 3, 29; S 3, 6, 8, 15, 17, 23 = 39
26. Foraminifera : A 1, 15, 19, 23, 24, 25, 31, 34, 56, 66, 67, 88; B 23, 24; C 43, 44, 45; D 22, 29, 44, 46; E 12; G 26 to 47, 70, 71, 72, 74, 77, 78; H 8, 12, 13, 17, 24, 26, 57; K 7; L 3, 9, 10, 11, 13, 14, 16, 48; M 12, 13, 32, 44 to 47; N 26, 30, 31, 34, 35, 36, 40 to 43; O 8, 9, 10, 13 to 19; P 18, 19, 43, 45, 102, 104, 129, 162, 169, 170, 171; Q 2, 3; R 1, 35 to 39, 42, 43; S 14, 21, 22, 32, 40, 53, 54, 55, 76; T 20, 21, 24, 30, 38, 53, 54, 55, 59, 62, 63; U 5; V 1, 3 = 137
27. Fungi : C 124, 125; i 10 = 3
28. Gastropoda : A 28, 58, 68; B 10, 41; C 25, 70, 71, 72, 97, 99; D 1; E 35; G 81, 82; H 56; L 47, 54 to 57; M 23, 29, 58; N 5, 27; P 20 to 24, 85 to 91, 103, 110, 111, 113, 114, 128, 164; R 10; T 16, 60, 61, 65, 66; V 15, 16; W 8 = 53
29. Ginkogales : G 19; P 175 = 2
30. Giraffidae : C 7; G 20; H 53; P 143; S 27, 28; V 12 = 7
31. Gymnospermae : D 31, 32; G 18; P 64 = 4
32. Insecta : G 63; K 5; M 26; P 140, 174 =5
33. Invertebrata : B 9; C 86, 94; M 4, 36; O 4; P 25; S 16, 43, 75; T 48 = 11
34. Labyrinthodontia : G 64; M 14, 35; P 49 = 4
35. Lamellibranchia : A 11, 30, 37, 50, 89 to 92; B 16, 43; C 17 to 20, 23, 60, 75, 76, 87, 90; D 8, 9, 18, 19, 25, 55; E 14, 15, 16, 36 to 43; G 65, 68, 69; H 3, 44; i 8, 16, 17; L 25, 26, 43, 49; M 18, 24, 30, 48, 56, 60, 61; O 36, 42; P 56, 58, 73, 82, 130, 153,, 165, 166, 167, 188; R 11; S 34, 35, 64, 90; T 13, 14, 44, 51; U 3; V2 = 79
36. Mammali a : A 41; D 14, 15, 17, 27; S 2, 87; X 1 = 8
37. NA (fossils for which the author could not find the fossil group) : A 12, 38, 93; B 11; C 8, 86; F 10; H 25, 55; L 12; P 132; R 12; T 1, 46; Y 1 = 15
38. NF (nanofossils) : C 37; M 42; Q 1 = 3
39. Ostracoda : A 16, 17, 18, 73; B 1, 2, 3, 45; C 10 to 15, 101, 107, 108, 111 to 123, 126 to 130; D 11, 12; E 31, 32, 33; F 8; G 58 to 61; H 22, 23, 49; L 30; M 7, 8, 49 to 54; N 39, 40; P 14, 15, 37, 40, 41, 42, 152; T 33; X 5, 6; Z 2 = 70
40. Palm : A 22; P 16, 17, 75 = 4

41. PF (palynofossil) : A 8, 45, 46, 47, 78, 94; C 2, 3, 24, 54, 66, 69, 74, 80, 81, 85, 88,, 96, 100, 103; D 5, 6, 7, 20, 21; E 1, F 1; G 1, 2, 3; i 6, 7; L 4, 38, 50; M 38, 39, 59; P 76, 77, 116, 117, 120, 150, 151, 159, 160; Q 4; R 45; S 1, 11, 41, 42, 70,, 78, 79; T 35, 36, 40, 49, 50, 57; V 6, 11, 13 = 65
42. Plantae : A 51, 55; C 98; D 26, 43; E 19; H 19, 54, 58, 59; i 15; L 17, 41, 44, 45, 46; M 5, 15, 31; N 29; O 6; P 121, 122, 173; R 4, 25; S 18; T 27 = 28
43. Porifera : C 6; i 4; L 37 = 3
44. Primates : B 37; G 16; H 47; K 23; P 27; R 5; S 26, 89 = 8
45. Proboscidea : A 57; D 16; E 2, 7, 8, 9, 10, 11; G 62; P 48; S 65, 66, 67; T 56 = 14
46. Pteridophyta : E 20; K 6; P 46, 83, 156, 176 to 180; R 7; S 12, 36; T 58 = 14
47. Pteridospermatophyta : D 28; G 5 to 14, G 48 to 57; R 21; S 57; T 28, 29; V 5 = 26
48. Reptilia : A 19; C 22, 34, 67; i 13; L 1, 34; M 3, 28; R 13; S 19; T 9, 32, 34; W 7 =15
49. Rhinocerotidae : A 6; B 34, 35; C 38, 61; G 4; R 22, 23, 24 = 9
50. Rodentia : C 73; H 60; K 1; N 28; R 27 = 5
51. Suidae : C 68; D 49; H 40; L 36, 39; P 148; S 91 to 96; T 18 = 13
52. TF (Trace fossils) : A 59, 69, 70, 85; B 13; C 41, 59, 95; D 33, 39, 40, 42; G 79, 80; H 16; i 1, 18; M 55; N 17; O 12; P 10, 57, 78 to 81, 95 to 101, 112, 118, 155, 161; R 26, 40, 41, 44, 46; S 29, 30; T 12, 22, 23, 31; Z 3 = 49
53. Trilibita : A 32, 33; C 5; D 3, 4, 41; E 3, 13; H 51; L 33, 58; O 31, 32, 33; P 7, 70, 74, 185; R 14 = 19
54. Vertebrata : A 7, 21, 52; B 38; C 30, 36; i 19; R 9 = 8

TOTAL = 1217

■■■■

X

LIST OF FOSSILIFEROUS FORMATIONS

S.No.	Formation	
	No.	Name
1	A 4	Agglomeratic Slate
2	A 6	Aida Stage
3	A 22	Anaipadi Formation
4	A 23	Anceps Beds
5	A 35	Ariyalur Group
6	A 42	Athleta Beds
7	B 3	Babia Stage
8	B 5	Badhaura Formation
9	B 8	Bagh Beds
10	B 10	Bahreli Formation
11	B 14	Baisakhi Formation
12	B 24	Barakar Formation
13	B 25	Baratang Group
14	B 26	Baripada Beds
15	B 31	Basal Muschelkalk
16	B 42	Bhadasar Formation
17	B 53	Bhuj Beds
18	B 58	Bijori Formation
19	B 71	Boulder Slate Formation
20	B 72	Broach (Bharuch) Formation
21	C 16	Chari Formation
22	C 26	Chhojan Member
23	C 27	Chhongtash Formation

S.No.	Formation	
	No.	Name
24	C 31	Chikkim Formation
25	C 33	Chilka Lake
26	C 45	Coastal Gondwana Formation
27	C 46	Conglomerate Bed, Narmada valley
28	C 47	Coral Limestone
29	C 56	Cuddalore Sandstone Formation
30	D 1	Dagshai Formation
31	D 8	Daonella Limestone
32	D 9	Daonella Shales
33	D 28	Dharangadhara Group
34	D 34	Dhosa Oolite
35	D 40	Dinod Formation
36	D 48	Dras Formation
37	D 49	Dras Volcanics
38	D 50	Drumgam Formation
39	D 51	Dubrajpur Formation
40	F 2	Fenestella Shale
41	G 1	Gaj Formation
42	G 2	Gajansar Beds
43	G 5	Gangamopteris Beds
44	G 6	Gangapur Formation
45	G 8	Garbyang Formation
46	G 10	Gechang Formation
47	G 15	Giumal Formation
48	G 22	Gondwana of Extrapeninsula
49	G 23	Gondwana Supergroup: (a) Lower Division (b) Upper Division
50	G 28	Grey Beds
51	G 31	Gund Formation
52	G 33	Gungri Formation
53	G 35	Gushal Formation
54	H 1	Habur Formation
55	H 4	Halobia Beds

S.No.	Formation	
	No.	**Name**
56	H 12	Hedenstroemia Beds
57	H 20	Hoti Member
58	H 21	Hundri Formation
59	i 5	Infra-trappean beds 1. Andhra Pradesh: a. Marepalli, b. Rajahmundry 2. Gujarat : a. Rahioli 3. Madhya Pradesh : a. Jabalpur 4. Maharashtra : a. Dongargaon, b. Pisdura, c. Umrer 5. Krishna-Godavari Basin : a. Chintalapalli-A, b. Elamanchili-A, c. Modi-A d. Narsapur Well-1, e. Palakollu-A
60	i 8	Inter-trappean Beds 1. Andhra Pradesh : a. Asifabadb, b. Rajahmundry, c. Naskald, d. Rangapur 2. Gujarat : a. Anjar, b. Bamanbor, c. Kora, d. Lakshmipur 3. Karnataka : a. Biligi, b. Gurmatkal, c. Upparhatti, d. Yellur 4. Madhya Pradesh : a. Kisalpuri, b. Mohgaon-Kal, c. Padwar/Ranipur, d. Mandla, e. Sagar 5. Maharashtra : a. Bombay (Mumbai), b. Nagpur 6. Rajasthan : a. Mamoni 7. Uttar Pradesh : a. Lalitpur 8. Krishna - Godavari Basin: a. Chintalapalli-A, b. Elamanchili-A, c. Modi-A, d. Narsapur Well-1, e. Palakollu-A, f. Razole-A 9. Rajmahal Basin 10. Chimalgutta 11. Mallangutta
61	J 22	Juvavites Beds
62	K 5	Kakdi Stage
63	K 9	Kalhal Formation
64	K 11	Kallakkudi Limestone Formation
65	K 12	Kallamedu Formation
66	K 13	Kallankurichi Formation
67	K 19	Kand/Kond Formation
68	K 21	Kankawati Series
69	K 23	Kanthkot Sandstone
70	K 26	Karai Shale Formation
71	K 38	Katrol Formation
72	K 42	Khalsi Formation
73	K 44	Khari Series

S.No.	Formation	
	No.	Name
74	K 51	Khunamuh Formation
75	K 55	Kioto Member
76	K 62	Kota Formation
77	K 64	Krol Formation
78	K 68	Kukti Formation
79	K 69	Kulakkalnattam Formation
80	K 70	Kuling Formation
81	K 73	Zanskar Area, J & K
82	L 8	Lamayuru Formation, Laddakh, J & K
83	L 13	Lapthal Member
84	L 24	Lipak Formation
85	L 29	Lower Katrol
86	L 31	Lower Muschelkalk
87	M 2	Macrocephalus Beds
88	M 4	Mahadek Formation
89	M 19	Mamal Formation
90	M 22	Manendragarh Marine Bed
91	M 28	Maravattur Clay Formation
92	M 39	Meekoceras Zone
93	M 40	Megalodon Limestone/Para Stage
94	M 45	Middle Katrol
95	M 49	Miliolite/Porbandar Stone
96	M 54	Miri Formation
97	M 57	Monotis Shales
98	M 66	Muth Quartzite/Formation
99	N 4	Nagri Formation
100	N 23	Niniyur Group/Formation
101	N 25	Nishatbagh Bed
102	N 26	Niti Limestone Member
103	N 27	Niti Member
104	O 6	Ophiceras Zone
105	O 7	Ordovician-Silurian rocks
106	O 8	Otoceras Zone
107	O 9	Ottakovil Formation

S.No.	Formation	
	No.	**Name**
108	P 1	Pachcham Basal Beds
109	P 2	Pachcham Coral Bed
110	P 4	Pachcham Shell Limestone
111	P 5	Pachmarhi Formation
112	P 21	Parahio Series
113	P 23	Pariwar Formation
114	P 31	Permian rocks
115	P 36	Piram Beds
116	P 37	Pleistocene Sediments of – a. Kurnool, Ghod, Manjra, Paravara & Godavari; b. Central Narbada (Up. Group), Palmer; c. Central Narbada (Lr. Group); d. Lr. Karewa, Pinjor, Up. Siwaliks
117	P 38	Po Formation
118	P 51	Productus Shale
119	Q 1	Qazil Langer Formation
120	Q 2	Quartzite Series
121	Q 3	Quaternary sediments of Andhra Pradesh
122	Q 4	Quaternary sediments of Central India
123	Q 5	Quaternary sediments of Gujarat Mainland
124	Q 8	Quaternary sediments of Maharashtra
125	Q 9	Quaternary sediments of Tamil Nadu
126	R 2	Raghvapuram Formation
127	R 9	Rajmahal Traps
128	R 12	Ramania Stage
129	R 13	Rambakot Member
130	R 23	Raniganj Formation
131	R 31	Rehmanni Beds
132	R 36	Rimkhim Member
133	S 4	Salooni Formation
134	S 5	Sancha Malla Group
135	S 13	Saser Brangsa/Shaksgam Formation
136	S 25	Seti Khola Formation
137	S 31	Shiala Formation
138	S 37	Sillakudi Formation

S.No.	Formation	
	No.	Name
139	S 49	Sivaganga Beds
140	S 50	Sivaganga Formation
141	S 51	Siwalik Group: a. Lower Division; b. Middle Division; c. Upper Division
142	S 61	Spiti Formation/Shale
143	S 62	Spiti Member
144	S 63	Sriperumbudur Formation
145	S 65	Subathu Formation/Group
146	S 73	Surma Group
147	S 79	Syringothyris Limestone
148	T 1	Tagling Limestone
149	T3	Takche/Manchap Formation
150	T 4	Takli Formation
151	T 5	Tal Group/Formation
152	T 6	Talchir Formation
153	T 14	Terani Plant Bed
154	T 18	Tertiary of Gujarat
155	T 20	Tethyan Basin
156	T 23	Thango Formation
157	T 34	Tirupati Formation
158	T 40	Triassic of NW Himalaya
159	T 44	Trichinopoly Group
160	T 45	Trigonia Beds
161	T 47	Tropites Beds
162	U 2	Ukra Beds
163	U 5	Umaria Marine Bed
164	U 6	Umia Ammonite Bed
165	U 7	Umia Beds
166	U 8	Umia Formation
167	U 11	Umia Stage
168	U 14	Upper Katrol sandstone
169	U 15	Upper Katrol Shales
170	U 21	Uttatur Group
171	V 5	Variegated Formation

S.No.	Formation	
	No.	**Name**
172	V 7	Vemavaram Formation
173	W 1	Wadhwan Formation
174	W 2	Waior Stage
175	Y 2	Yamne Formation
176	Y 5	Yerrapalli Formation
177	Z 4	Zewan Formation

■■■■